Pushing The Right Button

Editors: Allison Hrip, Aurora Corialis Publishing and Corlette Deveaux

Copy Editor: Corlette Deveaux, Alexis Deveaux

Design and Composition: Vern Cameron, RAGEHouse Creative

Publisher: Senior Living Consultants

Address: Senior Living Consultants, 254 Chapman Rd, Ste 208, #7500 Newark, DE 19702

Phone Number: +1 954.610.3732

First Published: 2024

© July 2024
No part of this publication may be reproduced, stored in a retrieval system, or transmitted in any form or by any means, including electronic or photocopying, without the prior permission of the publisher.

Designations used by companies to distinguish their products are often claimed as trademarks. All brand names and product names used in this book are trade names, service marks, trademarks, or registered trademarks of their respective owners.

Limit of Liability / Disclaimer

The content of this book is intended for learning and informational purposes only. The information contained in this book is not intended to provide legal advice or meant to replace any instructions or advice given by the manufacturer of each device.

The views expressed within this book are those of the author and do not necessarily reflect the views or opinions of any organization, manufacturer, employer, publisher, or individual. As evolving human beings, we recognize that our views may change over time, subsequently, the views of the author are subject to change. No guarantees are implied or expressed by the publisher to include any of the content in this volume.

Instructions provided in this book are based on the latest operating systems. There are hundreds of cell phones on the market, therefore, we have limited the instructions to the most popular cell phone(s) in each major category. Contact the manufacturer for specific information on other phones that are not provided in this book. The basic phone category instructions are provided for the TCL Classic Android 14 AOSP. The smartphone category instructions are provided for the Galaxy Android 14 and the iPhone 15 iOS. No instructions are provided for the feature phones.

Please note, instructions for the cell phones in this book are subject to change based on future operating system updates.

If you choose to follow the instructions outlined in this book, the author and publisher advise you to take full responsibility for your results. The author and publisher are not liable for any damages or negative consequences from the products, application, instructions, services, or preparation to any person reading or following information in this book.

Table of Contents

Prefix
Overview
Acknowledgments

CHAPTER 1
Introduction to Cell Phones 11
Cell Phones 12
Types of Cell Phones 12
Cell Phones Key Features 14
Purpose of Cell Phones 15
Advantages and Disadvantages of Various Cell
Phones Types 16
Selecting the Right Cell Phone 22
Chapter Questions 27

CHAPTER 2
Making and Receiving Calls on Cell Phones 29
Making Phone Calls 30
    TCL Flip Phones 31
    iPhones 40
    Galaxy Phones 50
Receiving Calls 60
    TCL Flip Phones 61
    iPhones 69
    Galaxy Phones 77
Chapter Questions 85

CHAPTER 3
Sending Text Messages 87
Composing and Sending Text Messages 88
    TCL Flip Phones 89
    iPhones 98
    Galaxy Phones 107
Chapter Questions 116

CHAPTER 4
Taking Pictures, Selfies, and Videos 119
    TCL Flip Phones 121
    iPhones 130
    Galaxy Phones 139
Chapter Questions 148
Bonus – QR Codes 149

CHAPTER 5
Cell Phone Security and Troubleshooting 151
Getting Familiar with Your Cell Phone's Security
Features 152
    TCL Flip Phones 153
    iPhones 157
    Galaxy Phones 160
Protecting Personal Data and Recognizing
Scams 163
Chapter Questions 165

CHAPTER 6
Real-Time Text (RTT) and Teletype (TTY) 167
    iPhones 168
    Galaxy Phones 169

CHAPTER 7
The Benefits of Artificial Intelligence (AI) and Voice
Activated Technology for Seniors 171
Artificial Intelligence Assistants 172
Hands-free Voice Assistants 174
Voice Activated Technology 176

SENIOR RESOURCES
Resource Links 179
Crossword Puzzle 180
Word Scramble 181
Answer Key 182

Conclusion 183
About The Author 184

Preface

Life is an ongoing journey of learning and experiencing moments that shape our memories. Just as we exercise our bodies to stay fit, it's essential to keep our minds sharp through continuous learning. Embracing technology is one way to continue learning.

The world is constantly changing, and staying open to learning allows us to adapt to these changes. From the obsolescence of encyclopedias to the rise of electric cars and the convenience of online shopping, technology evolves rapidly.

Reflecting on the difficulties I faced teaching my late mother, Eunice Deveaux, how to use a cell phone, emphasized the need to help seniors adapt to modern technology. This book is dedicated to my mother, and those like her, challenged by the very technology they need to stay connected and keep their minds sharp. Learning the basics of using a cell phone could increase their daily communication and enjoyment.

Seniors represent a significant portion of the population, and empowering them with technological skills enhances their independence and enjoyment of life. Let's come together to support seniors in learning the fundamentals of cell phone use and boost their confidence in navigating the digital world.

Thank you,

Corlette Deveaux, MBA
Author and Senior Advocate
Senior Living Consultant
www.SeniorLC.com

Overview

Pushing The Right Button is a comprehensive guide tailored for seniors who may feel apprehensive about technology, particularly cell phones. This engaging book takes readers on a journey from understanding the various types of cell phones available to mastering their myriad functions for enhanced communication and daily enjoyment.

The book begins with demystifying the landscape of cell phones, providing clear explanations of types of cell phones and their features. It lays a solid foundation by helping seniors understand the basics, such as making and receiving calls, sending text messages, and more.

As the reader progresses, *Pushing The Right Button* dives deeper into the expansive capabilities of cell phones, unlocking their potential to connect seniors with loved ones, friends, and the wider world. It explores features like taking pictures, selfies, and videos, empowering seniors to stay connected and engaged in today's digital age.

Moreover, the book highlights protecting personal data and recognizing scams. It emphasizes how cell phones can foster independence by providing access to essential services like emergency assistance at the touch of a button.

Throughout *Pushing The Right Button*, the tone is supportive and encouraging, recognizing that everyone learns at their own pace. The book provides practical insights, easy-to-follow instructions, troubleshooting tips, and benefits of Artificial Intelligence to help seniors overcome challenges they may encounter along the way. And, for interactive learning, there are questions at the end of the instructional chapters.

Ultimately, *Pushing The Right Button* empowers seniors to embrace technology with confidence, enabling them to harness the full potential of their cell phones for communication, entertainment, and convenience. With its accessible approach and wealth of information, this book serves as a valuable resource for seniors looking to navigate the digital world with ease and enjoyment.

Acknowledgments

This book is devoted to my strong, compassionate, caring and phenomenal mother, Eunice Deveaux. Simply thank you for being you!

With gratitude and humility, I give thanks for the blessings bestowed upon me by my Lord and Savior.

I am deeply and humbly grateful to everyone who has supported me throughout my life's journey—you are truly appreciated.

To my beloved husband, Scott, who has always been my rock and my biggest advocate. Your unwavering commitment to our family and to me means more than words can express. Thank you for your endless love and support.

To my two biological children, Angelo and Alexis, you are my legacy and greatest gift from God. I am so proud of you both. Thank you for being my support system and rock through our journey together as DEVEAUXs. To my bonus children, Alexandra and Zachary, you both are amazing and I love you dearly. Keep growing and soaring. We are all proud of you.

To my cherished sisters, Krista, Vonette and Cynara, who provided input and encouragement through this journey... Thank you. Your unwavering support, and unconditional love are felt and reciprocated. I love you both more than words can say.

True friends are hard to find, and I feel blessed to have lifelong friends who show support and love reciprocally. To Michelle, Renae, and Denise I love our friendship and sisterhood.

To my dear friend, Jersey Connie, what can I say but thank you, thank you! You designed my book cover and spent hours on the phone with my team to help publish my first book. You are amazing, and I appreciate you tons.

To my friend, colleague and graphic design guru, Vern Cameron—you are a rock star! Your patience, diligence, and creativity have been a blessing throughout this long, arduous process. I am thrilled this is done so your eyes can finally stop twitching. Yippee!

To my manager and faithful colleague, Marta, your loyalty and dedication are sincerely appreciated. Thank you!

To my loving family members, including my Sister Twin, thank you for continued support, strength and love throughout my life and our journey together. Family is everything.

"Every day is a chance to learn and grow. Embrace the journey and let your wisdom shine brighter with each new discovery."

With love and gratitude,
Corlette Deveaux

CHAPTER 1

Introduction to Cell Phones

Familiarizing Seniors with Different Types of Cell Phones and Providing Information to Help Them Choose the Right One

Cell Phones

A cell phone is simply a telephone that doesn't need a landline connection. Users can make and receive phone calls, and some cell phones also offer the ability to send text messages, take pictures, browse the Internet, engage in social media activities, and more.

Types of Cell Phones

There are three major categories of cell phones:

- **Basic Phones** – instructions will be provided for the TCL Flip Phone. It is one of the most popular basic phones on the market sold by most of the top carriers. Some of the icons on the phone may vary from carrier to carrier.

- **Feature Phones** – while these phones are on the market, instructions are not provided for these phones in this book.

- **Smartphones** – instructions will be provided for the Apple iPhone using the operating system known as iOS and the Samsung Galaxy operating on the AOSP system.

Instructions subject to change based on future operating systems.

Basic Phones

These are simple cell phones that are primarily used for making calls and sending text messages. They lack advanced features like Internet access or application (app) support. Basic cell phones are often chosen for their long battery life and durability.

Feature Phones

Feature phones are less advanced than smartphones but still offer more functionality than basic cell phones. They typically have physical keypads and are used primarily for calling and texting. Some may have basic web browsing and app support.

Smartphones

Smartphones are the most common type of cell phone. Smartphones offer a wide range of features beyond basic calling and texting, including Internet browsing, app support, email, GPS navigation, and so much more. They often run on operating systems like iOS (Apple) or AOSP (Samsung).

Cell Phones Key Features

Basic Phones

- Voice Calls
- Text Messaging
- Long Battery Life
- Durability
- Affordability
- Simplified User Interface

Feature Phones

- Voice Calls
- Text Messaging
- Basic Internet Access
- Camera
- Media Playback
- Expandable Storage

Smartphones

- Advanced Operating System
- Touch Screen Display
- Internet Connectivity
- Multimedia Capabilities
- GPS Navigation
- Social Media Integration
- Email
- Virtual Assistants

Purpose of Cell Phones

Basic Phones

- The primary purpose of basic phones is to provide a reliable means of communication for voice calls and text messaging.
- They are often chosen by individuals who want a simple and cost-effective cell phone without the advanced features and complexities of smartphones.

Feature Phones

- Feature phones bridge the gap between basic phones and smartphones.
- They are suitable for users who want more than just voice calls and text messaging but don't need the full range of features and capabilities offered by smartphones.
- Feature phones are often chosen for their affordability and simplicity, making them popular choices in regions where smartphones may be less accessible or affordable.

Smartphones

- The primary purpose of smartphones is to provide users with access to information, entertainment, communication, and productivity tools all on one device.
- Smartphones are suitable for a broad audience, from individuals seeking advanced communication options to those interested in multimedia, productivity, and gaming.
- They are often used for work, entertainment, socializing, and staying connected in today's digital world.

Advantages and Disadvantages of Various Cell Phone Types

Advantages of Various Cell Phone Types

Basic Phones

1. **Reliability:** Basic phones are known for their reliability in making and receiving voice calls, especially in areas with weak network coverage.

2. **Long Battery Life:** They typically have a longer battery life compared to smartphones, often lasting several days on a single charge.

3. **Durability:** Basic phones are often more rugged and durable, making them suitable for outdoor activities and rough handling.

4. **Affordability:** They are generally more affordable, both in terms of the device cost and monthly service fees.

5. **Simplicity:** The straightforward user interface is easy to navigate, making it ideal for those who prefer simplicity and are not tech-savvy.

6. **Emergency Use:** Basic phones are commonly used as emergency backup devices due to their long battery life and reliability.

Advantages of Various Cell Phone Types

Feature Phones

1. **Enhanced Features:** Feature phones offer more features than basic phones, including better camera quality, basic Internet access, and media playback.

2. **Affordability:** They are generally more affordable than smartphones, making them accessible to a broader range of users.

3. **Physical Keypad:** Feature phones often have physical keypads, which some users find more comfortable for typing.

4. **Longer Battery Life:** Compared to smartphones, feature phones often have a longer battery life.

5. **Simplicity:** Feature phones strike a balance between basic phones and smartphones, offering more functionality without overwhelming users.

Advantages of Various Cell Phone Types

Smartphones

1. **Versatility:** Smartphones offer the most extensive range of features and capabilities, from apps to multimedia to productivity tools.

2. **High-Performance:** They have powerful processors, ample memory, and high-quality displays, making them suitable for various tasks.

3. **App Ecosystem:** Smartphones have access to a vast library of apps for almost any purpose, including communication, entertainment, and productivity.

4. **Internet Access:** They provide fast and convenient Internet access, including browsing, email, and social media.

5. **Multimedia:** Smartphones excel in photography, video recording, music playback, and streaming services.

6. **GPS and Navigation:** Built-in GPS is valuable for navigation and location-based services.

7. **Continuous Updates:** Smartphones typically receive regular software updates and security patches.

8. **Real-Time Text (RTT):** This is a new feature on some smartphones. RTT is helpful for those who have speech or hearing impairments. RTT allows users to carry on conversational text via a phone call.

Disadvantages of Various Cell Phone Types

Basic Phones

1. **Limited Features:** They lack advanced features like Internet access, app support, and high-quality cameras.

2. **Limited Communication Options:** Basic phones are primarily designed for voice calls and text messaging, limiting communication options.

3. **Limited Media Playback:** They may not support multimedia playback or advanced media features.

Feature Phones

1. **Limited App Support:** While they can run some basic apps, feature phones have a limited app ecosystem compared to smartphones.

2. **Limited Internet Access:** Internet access on feature phones is typically slow and limited to basic browsing.

3. **Lower Performance:** They have less processing power and memory than smartphones, which can affect performance.

Disadvantages of Various Cell Phone Types

Smartphones

1. **Complexity:** The sheer number of features and options can be overwhelming for some users, especially those unfamiliar with technology.

2. **Cost:** Smartphones tend to be more expensive both upfront and in terms of monthly service fees.

3. **Battery Life:** Many smartphones have shorter battery lives compared to basic and feature phones, often requiring daily charging.

4. **Fragility:** They are generally more delicate and prone to damage from drops and accidents.

5. **Privacy and Security Concerns:** Smartphones may be more susceptible to privacy and security risks, including data breaches and malware.

Selecting the Right Cell Phone

Selecting The Right Cell Phone

Selecting the right cell phone involves considering various factors and preferences. Here's a step-by-step guide to help make an informed decision:

1 Identify Needs

Determine your primary usage requirements. Are you looking for a communication device, a multimedia and entertainment hub, or a productivity tool? Consider your daily activities, such as work, hobbies, and social interactions to understand how you will use the phone.

2 Set a Budget

Determine how much you're willing to spend on the phone and any associated monthly costs (e.g., data plan, apps, accessories). Keep in mind that different phone types come with varying price ranges, with basic phones generally being the most budget-friendly.

3 Consider the Following Key Features

a. Communication Needs:
If you primarily need a phone for calls and basic texting, a basic or feature phone might suffice. For more advanced communication needs like email, social media, and messaging apps, consider a smartphone.

b. Multimedia and Entertainment:
If you enjoy photography, video streaming, music, or gaming, prioritize smartphones with good camera quality, vibrant displays, and sufficient processing power.

c. Productivity:
If you plan to use your phone for work or productivity tasks, consider a smartphone with a robust app ecosystem, document editing capabilities, and compatibility with work-related apps.

d. Internet Access:
Reliable and fast Internet access is essential for browsing, streaming, and online services. Smartphones excel in this area.

e. Battery Life:
If long battery life is crucial for your lifestyle (e.g., frequent travel, outdoor activities), look for phones known for their endurance or consider a basic phone.

f. Durability:
If you have a physically demanding job or an active lifestyle, rugged phones or protective cases may be worth considering.

4 Consider the Operating System

Decide whether you prefer iOS (Apple) or AOSP (Android). Each has its own ecosystem of apps, features, and user experiences. Consider how familiar you are with the chosen operating system, as this can impact your ease of use.

5 Consider the Size and Form Factors

Choose a phone size that feels comfortable to hold and use. Some individuals prefer compact phones, while others opt for larger screens. Consider whether you want a traditional candy bar-style phone, a slider, a flip phone, or a foldable device.

6 Do Research and Read Reviews

Read reviews and watch videos to learn about the pros and cons of specific phone models. Look for user feedback to understand real-world experiences with the phone.

7 Visit Retail Stores

Whenever possible, visit a physical store to hold and test the phone to see how it feels in your hand and how user-friendly it is.

8 Consider Future Needs

Think about your phone's longevity. Will it meet your needs for the foreseeable future, or do you expect your requirements to change soon?

9 Consider Various Factors

Consider factors like software updates and support, as these can impact your phone's longevity.

10 Compare Options

Create a shortlist of phones that meet your criteria and compare them based on features, price, and user feedback.

11 Make a Decision

Once you've gathered all the necessary information and weighed your priorities, make an informed decision based on your needs and budget.

Remember that there's no one-size-fits-all answer when it comes to choosing a phone. What's right for one person may not be suitable for another. Take your time, do your research, and prioritize the features and capabilities that matter most to you.

Chapter Questions

1. What are the types of cell phones covered in this section? (Check all that apply)

    ☐ Basic Phone
    ☐ Rotary Phone
    ☐ Smartphone
    ☐ Soft Phone
    ☐ Feature Phone

2. If you were to choose a phone that is easy to use and provides Internet connectivity, what type of cell phone would that be?

    ○ Basic Phone
    ○ Feature Phone
    ○ Smartphone

3. Why would you choose a smartphone over a feature phone?

    ○ Versatility
    ○ High Performance
    ○ Internet Access
    ○ Continuous Update
    ○ All of the Above

ANSWERS: 1. Basic Phone, Feature Phone, Smartphone | 2. Feature Phone (because they are less advanced than smartphones but still offer more functionality than basic cell phones) | 3. All of the Above

CHAPTER 2

Making and Receiving Calls on Cell Phones

Teaching Seniors the Basic Uses of Cell Phones

Making Phone Calls: Step-By-Step Instructions for Dialing Numbers, Using Contacts, and Making Emergency Calls

Making Regular Phone Calls – TCL Phones

1 **Powering On Your Basic Phone**

- Press and hold **Power/Hang Up Button** until the phone powers on. Type in your PIN code if necessary. Once unlocked, the **Home Screen** is displayed.

- If you don't know your PIN code or if you have forgotten it, contact your service provider. Do not store your PIN code within your phone, instead store your PIN in a location that is accessible without using phone.

- Do not share your PIN with others.

Home Screen

Power/Hang Up Button

② Dialing a Number

- From the **Home Screen**, use the **Keypad** to dial the phone number.

- If you make a mistake, press **Back/Delete Key** to delete the incorrect digits.

3) Making a Call

- Once you've entered the number, press the **green Place/Answer Call Button**.
- Wait for the person you're calling to answer, and start your conversation.

Place/Answer Call Button

④ Ending a Call

- During a call, press **Right Menu Button** or **Power/Hang Up Button** to end a call.

Using Contacts – TCL Phones

5 **Accessing Your Contacts**

- Open the **Contacts** app by pressing **OK** from the Home screen.
- Select the **Contacts** app icon using the **Navigation Ring**.
- Press **OK** to access your contacts.

OK • — • Navigation Ring

TCL Flip Phone

Finding a Contact

- Scroll through your contact list using the **Navigation Ring**.
- Press **OK** to view a contact's details.
- Press **Right Menu Button** to access more contact's options.

 **Calling a Contact**

- In the contact's details screen, select the contact's number you want to call using the **Navigation Ring** and press the green button to **Place/Answer Call Button**.

Place/Answer Call Button

Navigation Ring

Making Emergency Calls – TCL Phones

 **Dialing Emergency Services**

- In any situation where you need immediate help, you can dial emergency services by following these steps:

- Dial emergency number (911 or your country's emergency number if you're outside the United States), and press **Place/Answer Call Button** to make the emergency call.

- Speak clearly and provide your location and the nature of the emergency to the operator.

- Emergency calls work without a SIM (subscriber identity module) card but still require network coverage.

Place/Answer Call Button

By following these step-by-step instructions, you should now be comfortable making phone calls on your basic phone, using your contacts, and making emergency calls when needed. Don't hesitate to seek assistance from family members or friends if you have any questions or encounter difficulties.

A Note About Feature Phones

In this book, no step-by-step instructions are provided for feature phones. While they may still be on the market, they have declined in popularity with smartphones being the most popular option for seniors.

NOTES

_____

_____

_____

_____

_____

_____

_____

_____

Making Regular Phone Calls – iPhones

1 ## Unlock Your iPhone

- Press the **Power** or **Side Button** on your smartphone to wake it up.

- If your phone has a security feature like a **PIN**, **Pattern ID**, **Touch ID** (fingerprint), or **Face ID** lock, unlock it by following the on-screen prompts.

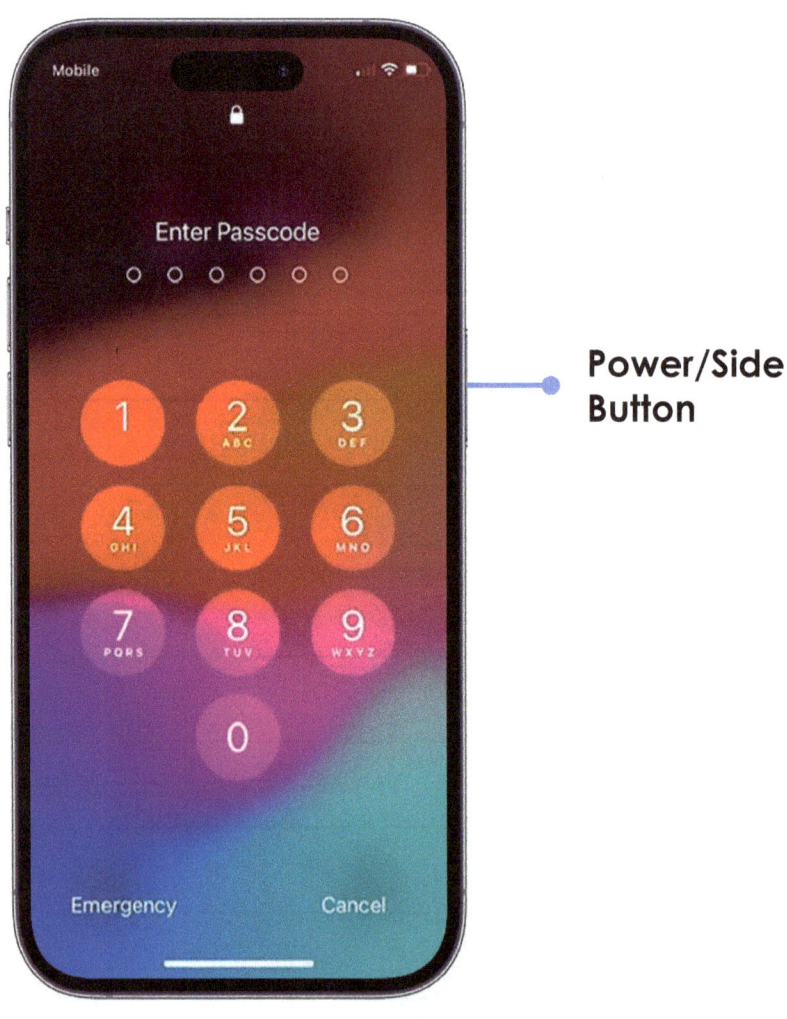

Power/Side Button

② Access the Phone App

- Look for the **Phone App Icon** on your home screen or in your **App Drawer** and tap it to open. It usually resembles an old-fashioned phone handset.

Phone Icon

App Drawer

3 Dialing a Number

- Tap the **Keypad Icon** within the phone app to access the dialer.
- Use your fingertip to tap the numbers on the **Keypad** to enter the phone number you wish to call.

iPhone

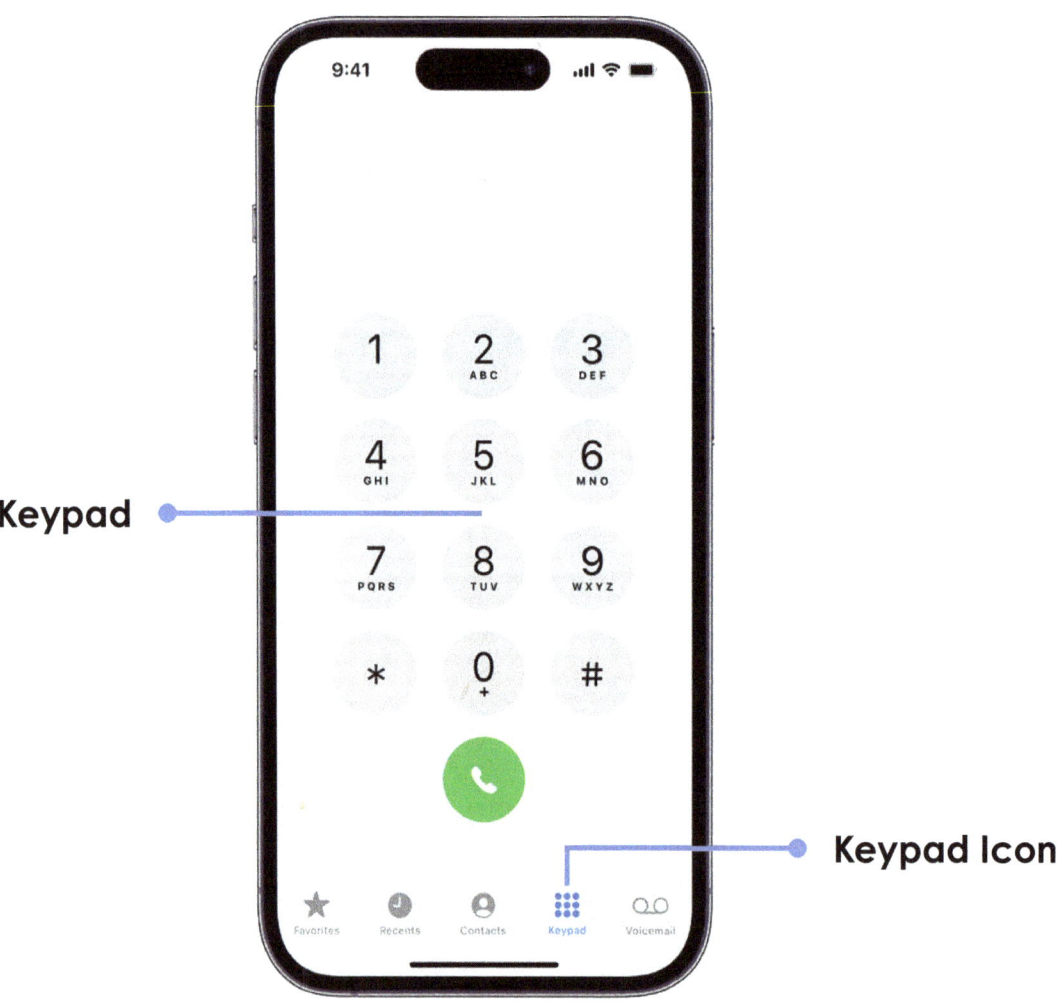

Keypad

Keypad Icon

4 Making a Call

- Once you've entered the number, tap the **green Call Button** to initiate the call.
- Wait for the person you're calling to answer, and start your conversation.

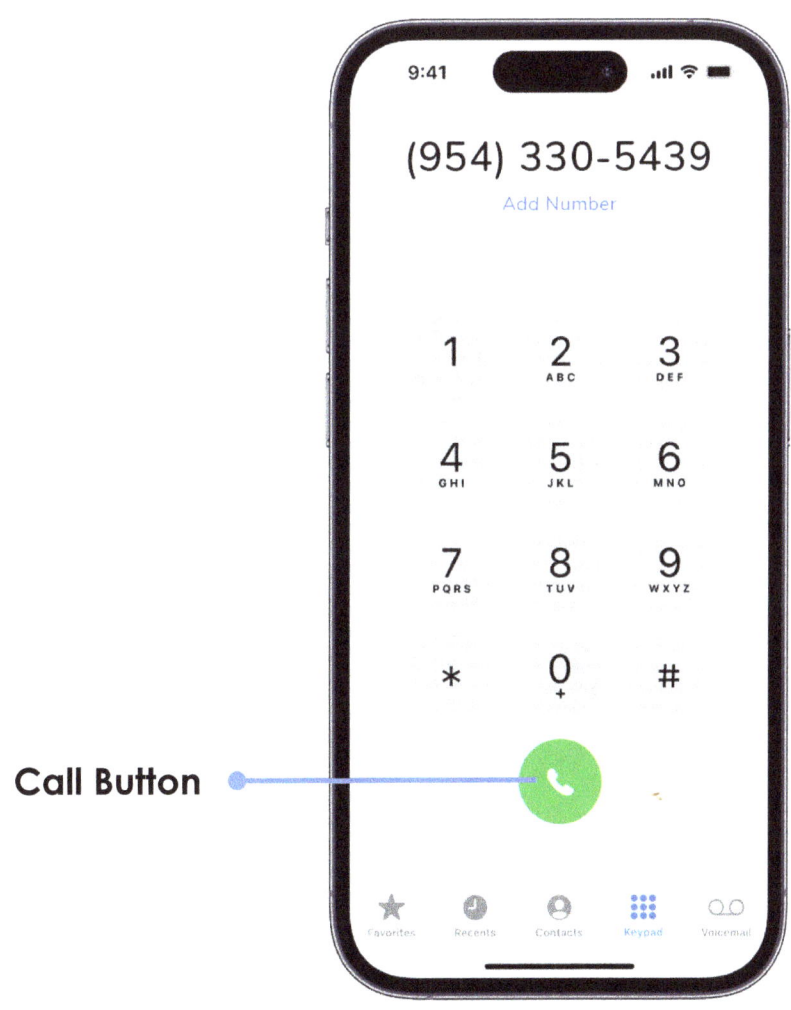

Call Button

iPhone

43

5 Ending a Call

- During a call, you'll see a **red End** or **Hang Up Button** on the screen. Tap it to end the call when you're finished.

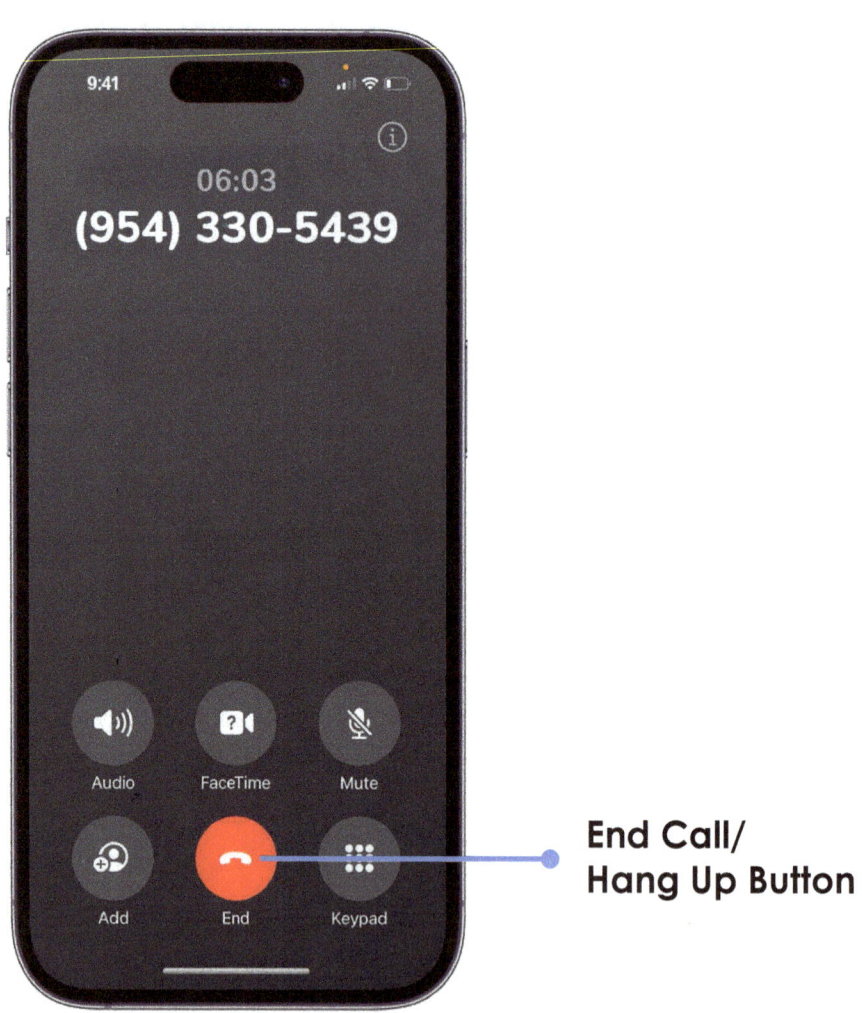

End Call/
Hang Up Button

Using Contacts – iPhones

Accessing Your Contacts

- Open the **Phone app**.
- Look for the **Contacts** or **Address Book Icon** within the app (usually represented by a person's silhouette).
- Tap it to access your **Contacts**.

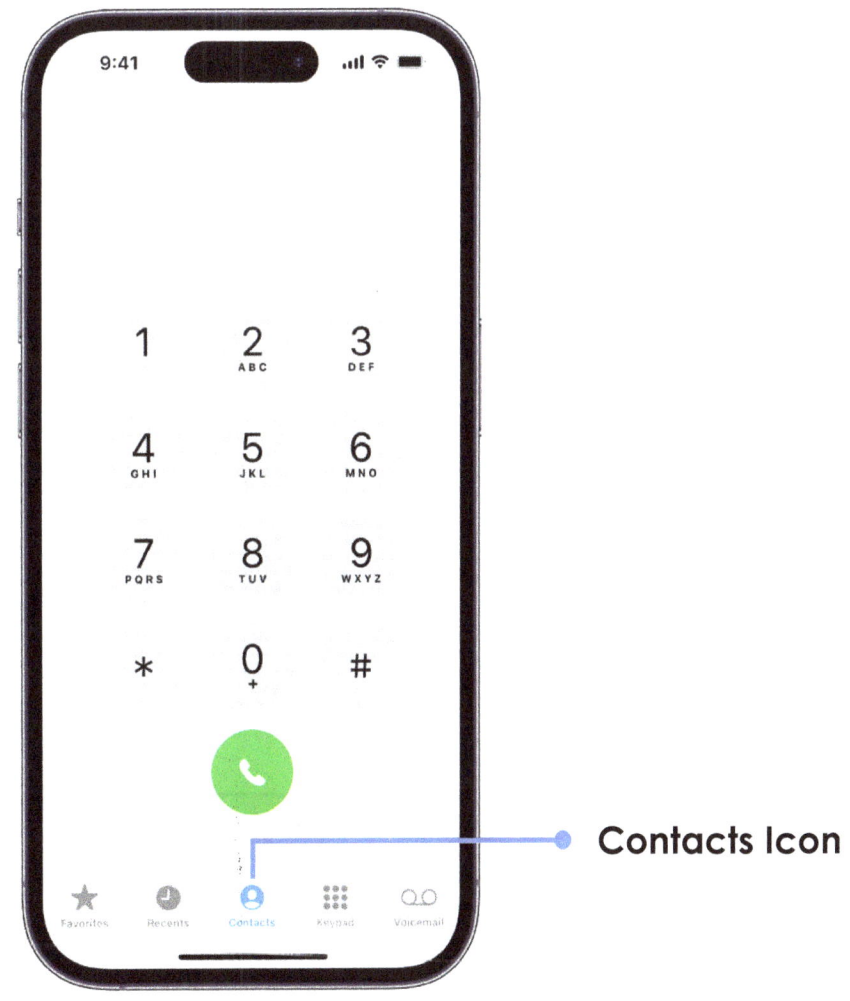

Contacts Icon

7 Finding a Contact

- Scroll through your **Contact List** by swiping up or down with your fingertip.

- Alternatively, use the **Search Bar** at the top of the screen to type the contact's name.

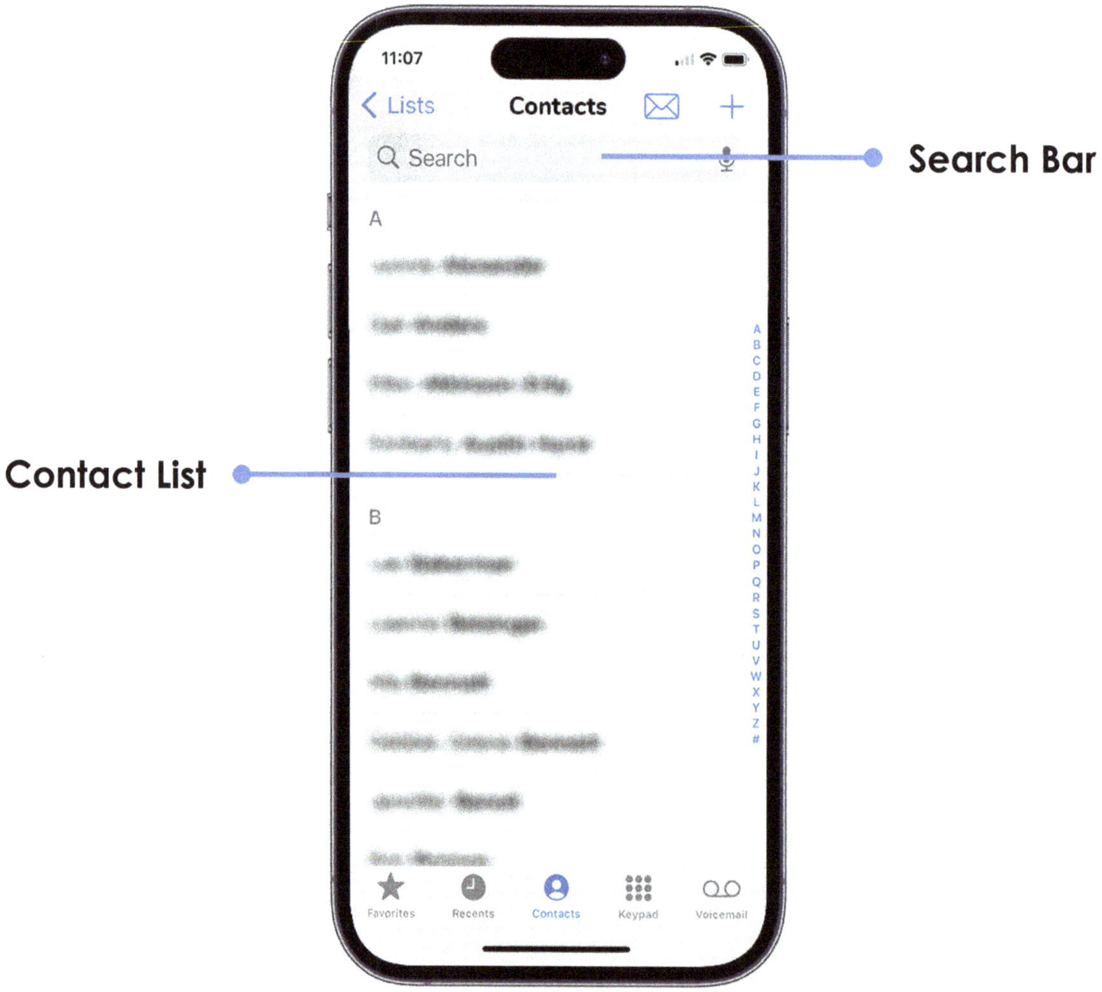

Search Bar

Contact List

iPhone

 **Calling a Contact**

- Once you've found the contact you want to call, tap on **the name**.

- This will open the **Contact's Details**. Look for a **Phone Icon** or **Call Button** and tap it to initiate the call or tap the phone number to place the call.

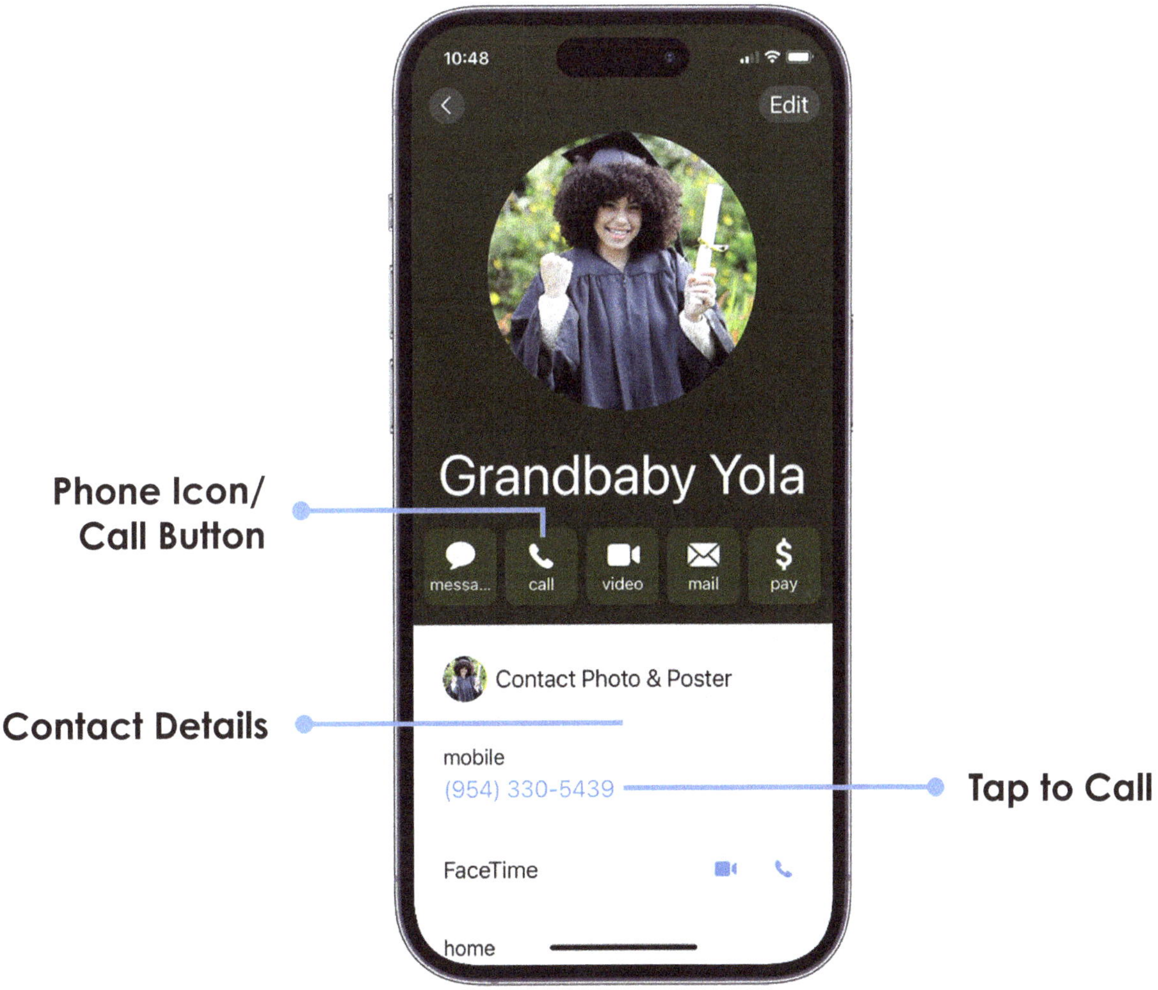

Phone Icon/Call Button

Contact Details

Tap to Call

Making Emergency Calls – iPhones

 **Dialing Emergency Services**

- In any situation where you need immediate help, you can dial emergency services by following these steps:

- On the lock screen, swipe up or press the **Emergency Button** (often depicted by a red SOS symbol).

- **Dial 911** (or your country's emergency number if you're outside the United States).

- Speak clearly and provide your location and the nature of the emergency to the operator.

Emergency Button

By following these step-by-step instructions, you should now be comfortable making phone calls on your iPhone (smartphone), using your contacts, and making emergency calls when needed. Don't hesitate to seek assistance from family members or friends if you have any questions or encounter difficulties.

NOTES

_____

_____

_____

_____

_____

_____

_____

_____

_____

_____

_____

_____

_____

Making Regular Phone Calls – Galaxy Phones

1 **Unlock Your Galaxy Phone**

- Press the **Power** or **Side Button** on your smartphone to wake it up.
- If your phone has a security feature like a **PIN**, **Pattern ID**, **Touch ID** (fingerprint), or **Face ID** lock, unlock it by following the on-screen prompts.

Power/Side Button

2 Access the Phone App

- Look for the **Phone App Icon** on your home screen or in your **App Drawer** and tap it to open. It usually resembles an old-fashioned phone handset.

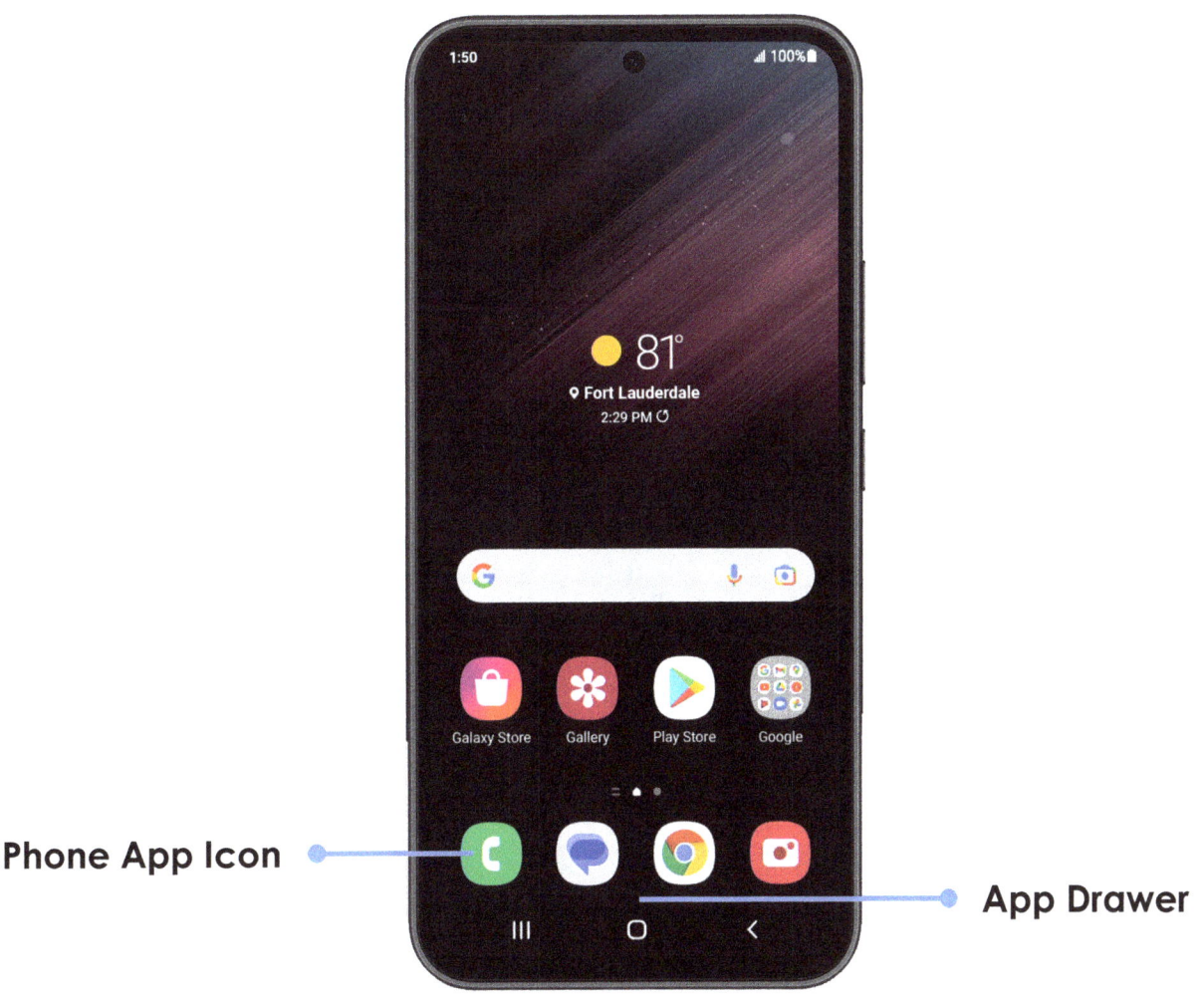

Phone App Icon

App Drawer

③ Dialing a Number

- Tap the **Keypad Icon** within the phone app to access the dialer.
- Use your fingertip to tap the numbers on the **Keypad** to enter the phone number you wish to call.

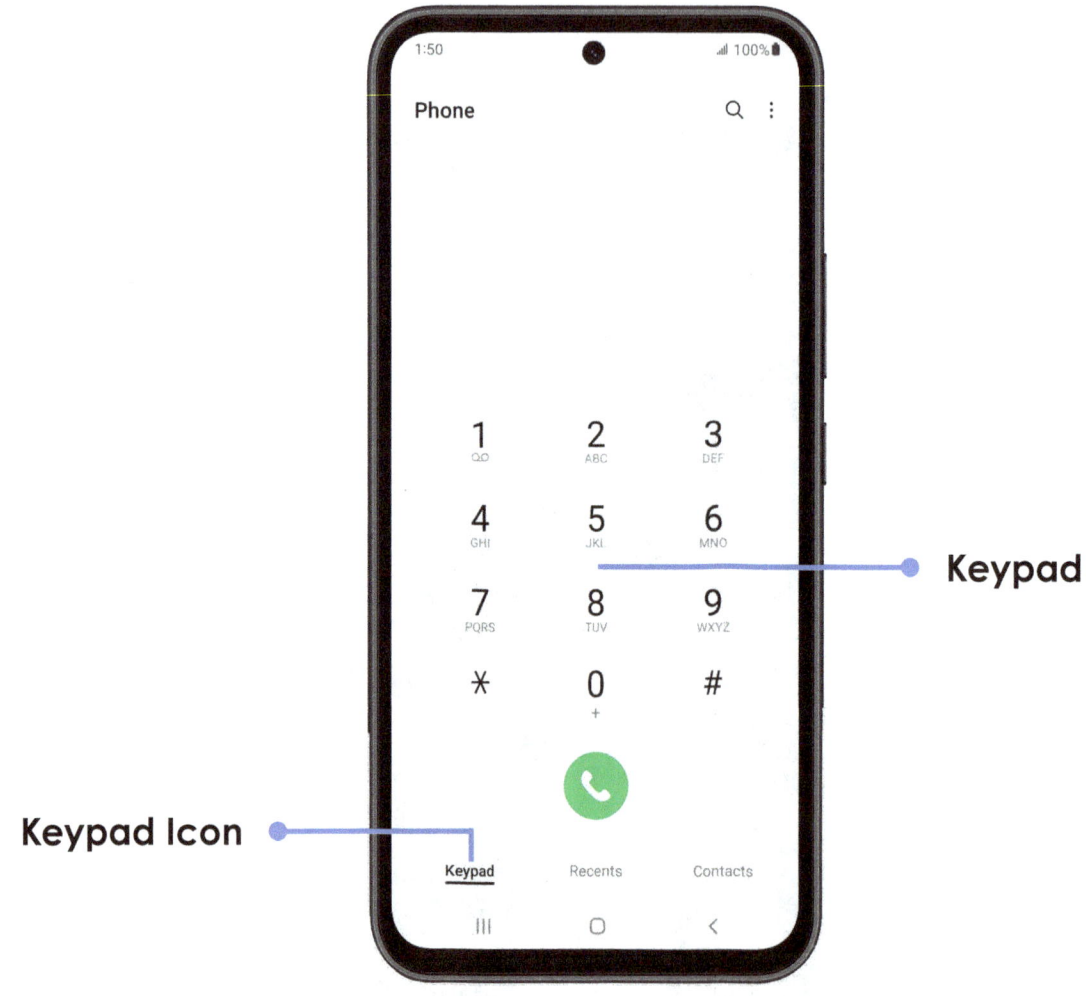

4 Making a Call

- Once you've entered the number, tap the **green Call Button** to initiate the call.//

- Wait for the person you're calling to answer, and start your conversation.

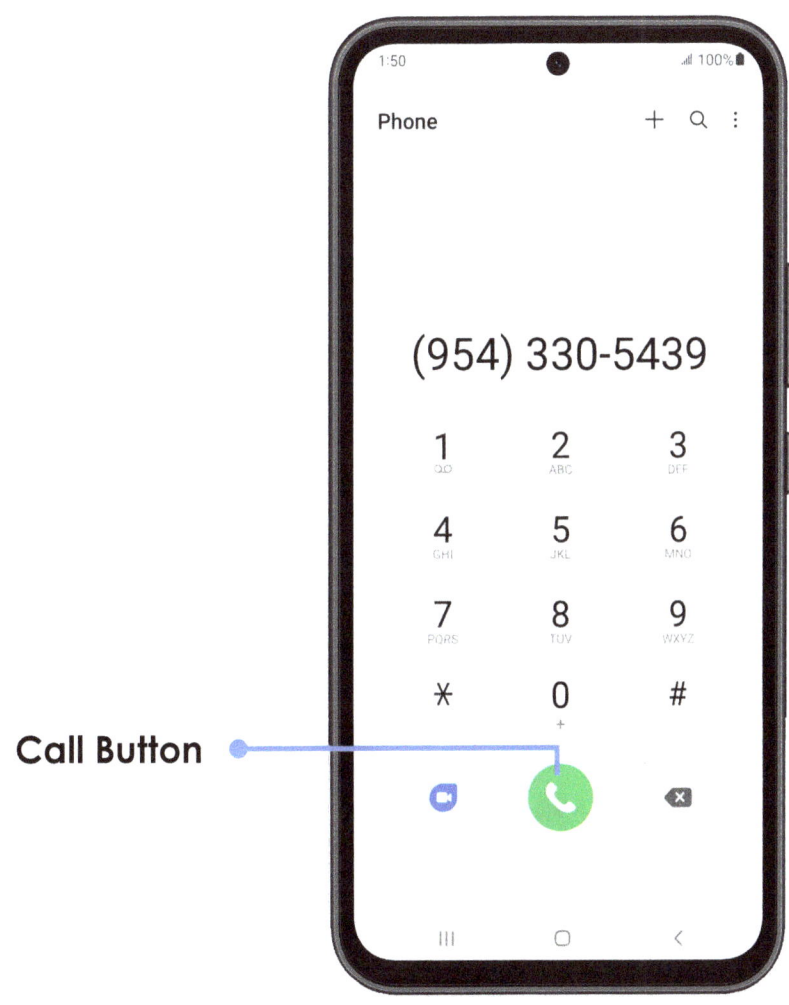

Call Button

⑤ Ending a Call

- During a call, you'll see a **red End Call** or **Hang Up Button** on the screen. Tap it to end the call when you're finished.

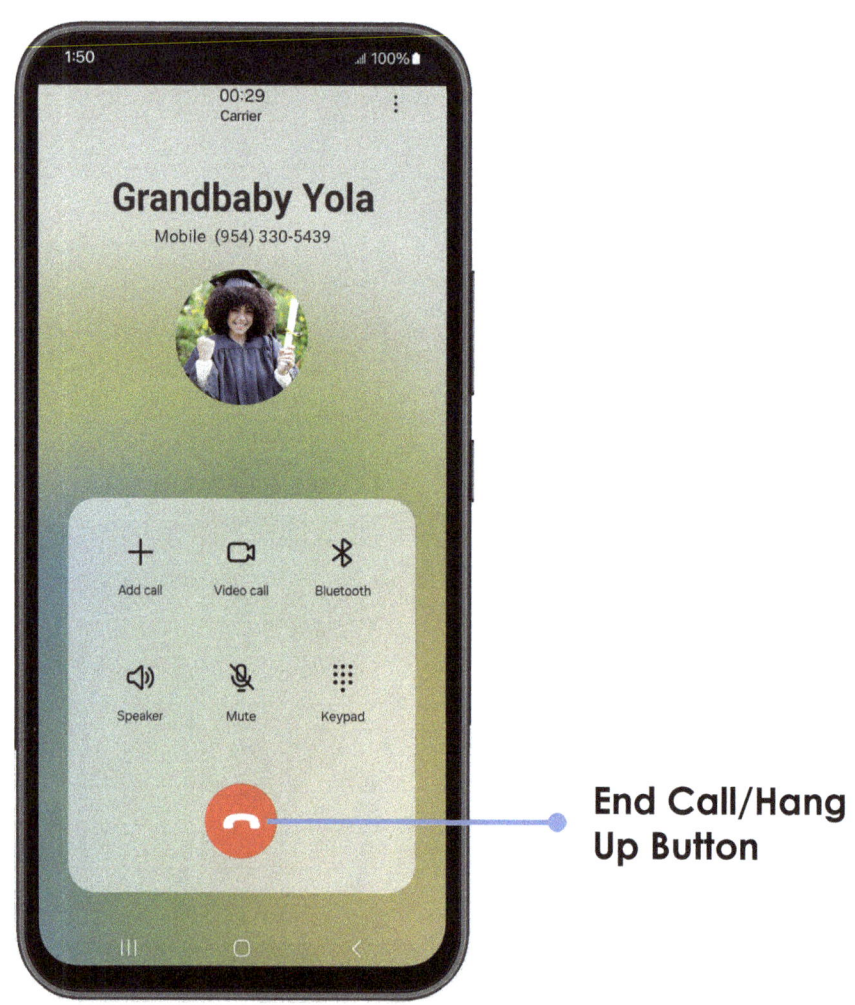

End Call/Hang Up Button

Using Contacts – Galaxy Phones

6 Accessing Your Contacts

- Open the **Phone app** .
- Look for the **Address Book** or **Contacts Icon** within the app.
- Tap it to access your contacts.

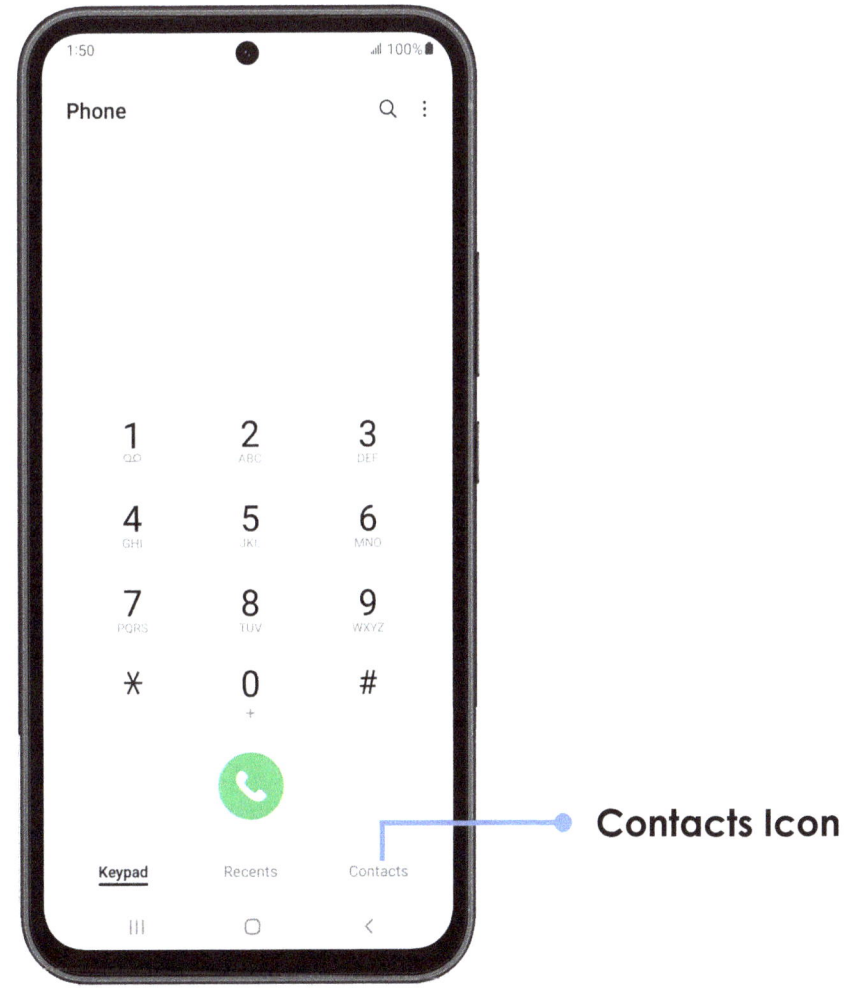

Contacts Icon

7 Finding a Contact

- Scroll through your **Contact List** by swiping up or down with your fingertip.

- Alternatively, use the **Search Icon**, usually represented by a magnifying glass, at the top of the screen to type the contact's name.

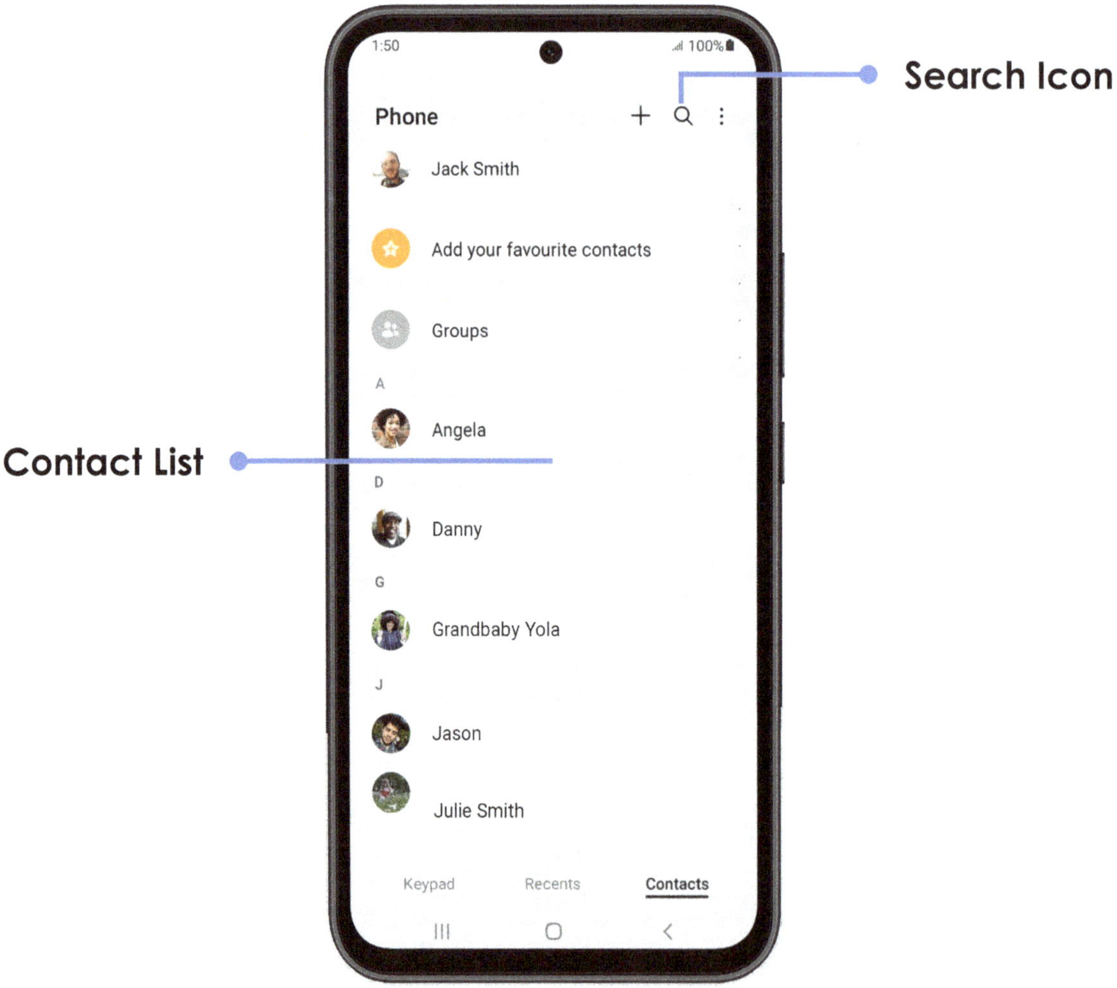

8 Calling a Contact

- Once you've found the contact you want to call, **tap the name**.
- This will expand the **contact's call options**. Look for a **green Phone Icon** or **Call Button** and tap it to initiate the call.

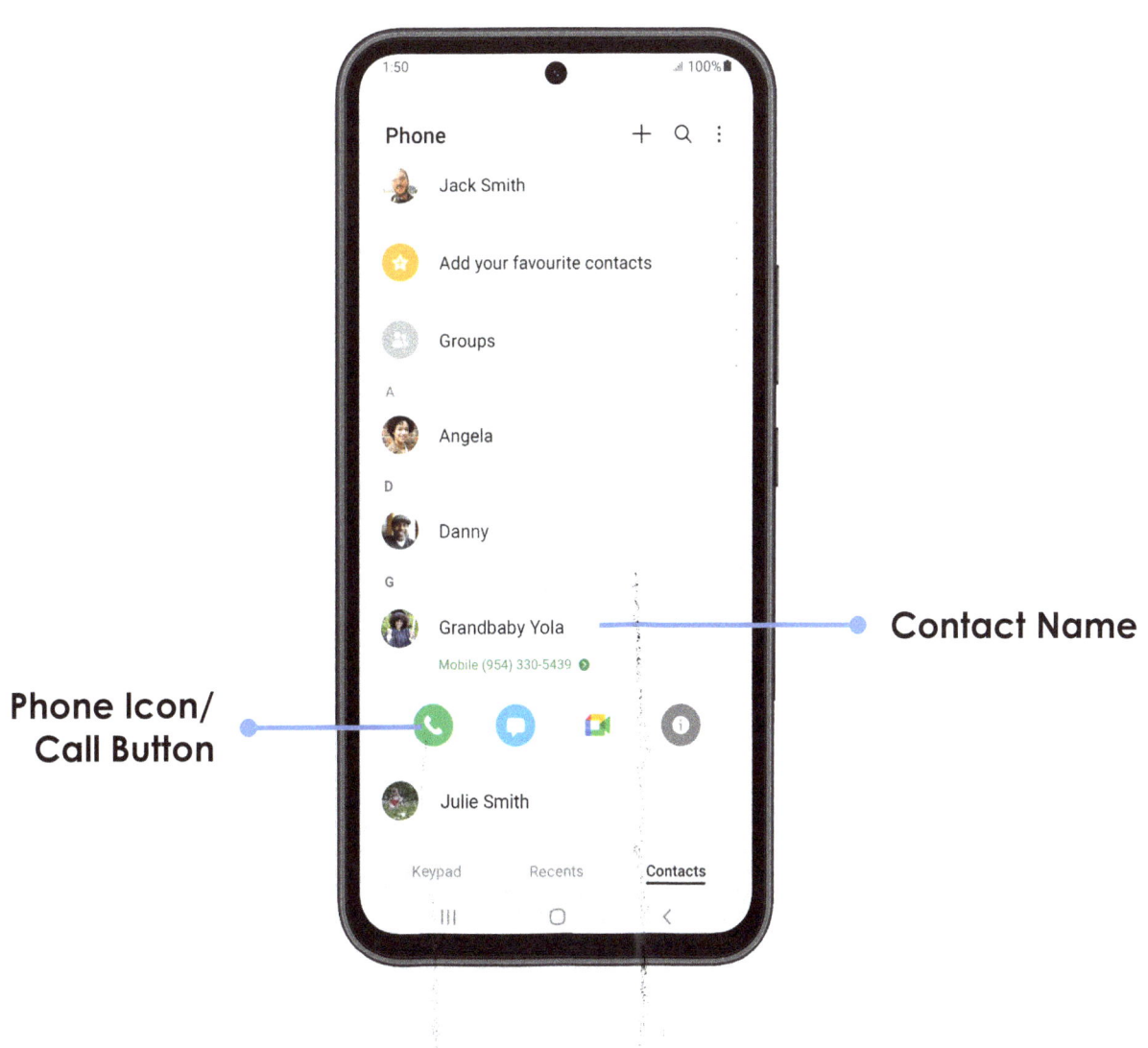

Contact Name

Phone Icon/ Call Button

Making Emergency Calls – Galaxy Phones

 **Dialing Emergency Services**

- In any situation where you need immediate help, you can dial emergency services by following these steps:

- On the lock screen, swipe up or press the **Emergency Call Button** (often depicted by a red SOS symbol).

- **Dial 911** (or your country's emergency number if you're outside the United States).

- Speak clearly and provide your location and the nature of the emergency to the operator.

Emergency Call Button

Keypad

By following these step-by-step instructions, you should now be comfortable making phone calls on your Samsung Galaxy (smartphone), using your contacts, and making emergency calls when needed. Don't hesitate to seek assistance from family members or friends if you have any questions or encounter difficulties.

NOTES

_____

_____

_____

_____

_____

_____

_____

_____

_____

_____

_____

_____

_____

Receiving Calls: How to Answer Incoming Calls, Reject Calls, and Use Call-Waiting Features

Answering Incoming Calls – TCL Phones

 **Recognizing an Incoming Call**

- When someone calls you, your basic phone will ring, and you'll see the caller's name or phone number displayed on your screen.

- You may also hear a ringing or vibrating sound.

Caller's Name/ Phone Number

② Answering a Call

- To answer the call, press the green **Place/Answer Call Button** or the **OK**. This will connect you to the caller.

- If your flip phone is closed, **open the flip** to answer.

OK

Place/Answer Call Button

3 Ending a Call

- When you're finished with the conversation, press the **red Power/Hang Up Button** to end call.

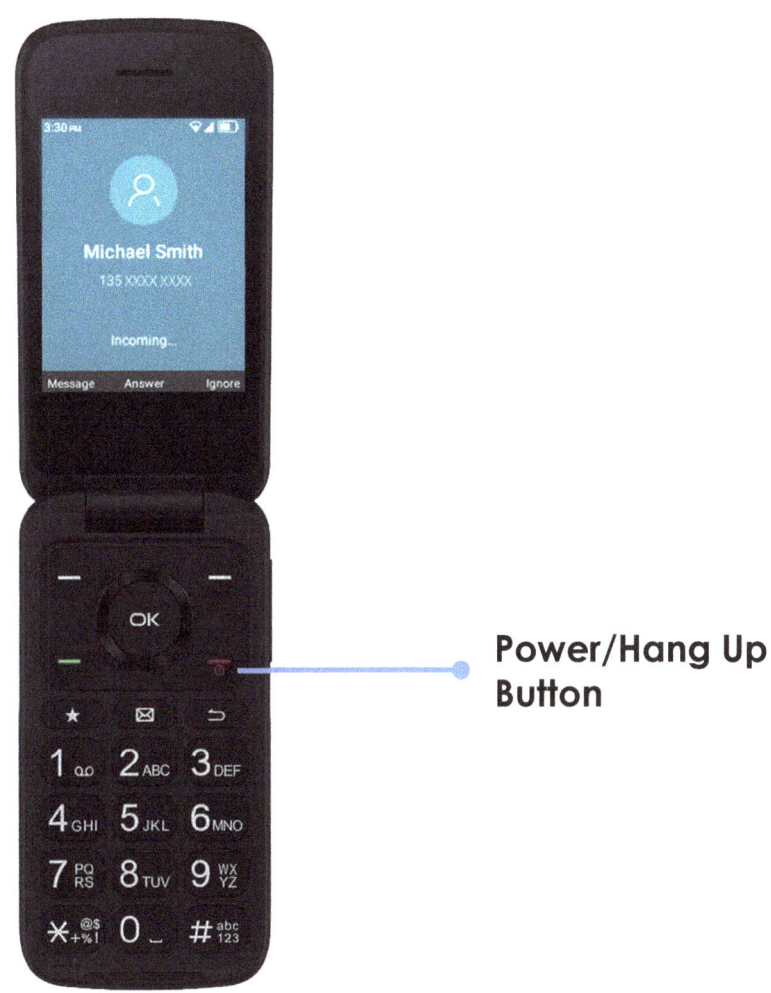

Power/Hang Up Button

Rejecting Incoming Calls – TCL Phones

 **Unrecognizable Incoming Call You Want to Reject**

- If you receive a call from a number you don't recognize or wish to reject, you can decline the call.

- When the call comes in, you'll see the onscreen options to **Message**, **Answer**, or **Ignore** on your screen.

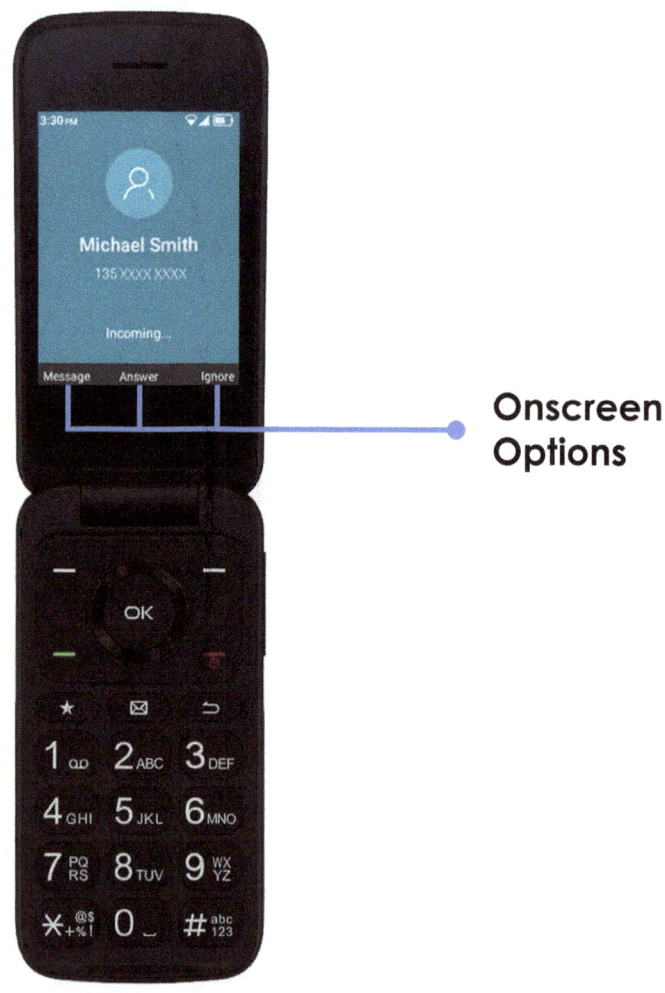

Onscreen Options

Rejecting a Call

- To reject the call, press the red **Power/Hang Up Button** or the **Right Menu Button** to ignore and decline.

- The caller will hear a busy tone and be sent to your voicemail.

- Or press the **Left Menu Button** to reject the call by sending a preset message.

Left Menu Button

Right Menu Button
Power/Hang Up Button

TCL Flip Phone

Using Call-Waiting Features – TCL Phones

Managing Call-Waiting

- When you're on a call and a new call comes in, you'll hear a call-waiting tone and see a name and/or number on the display.

You can access In-call options:

- End the current call by pressing the **Power/Hang Up Button** and answer the new call by pressing the **Place/Answer Call Button**.

- Put the current call on hold by pressing the **Right Menu Button** to select **Options**. Use the **Navigation Ring** to highlight **Hold Call**, press **OK**, then answer the new call.

- Ignore the new call and let it go to voicemail.

 **Switching Between Calls**

- If you're on a call and want to switch between active calls (e.g., to alternate between callers), press the **Right Menu Button** to select **Options**.

- Press the **Navigation Ring** to highlight **Swap Call**, then press **OK** to return to the call on hold.

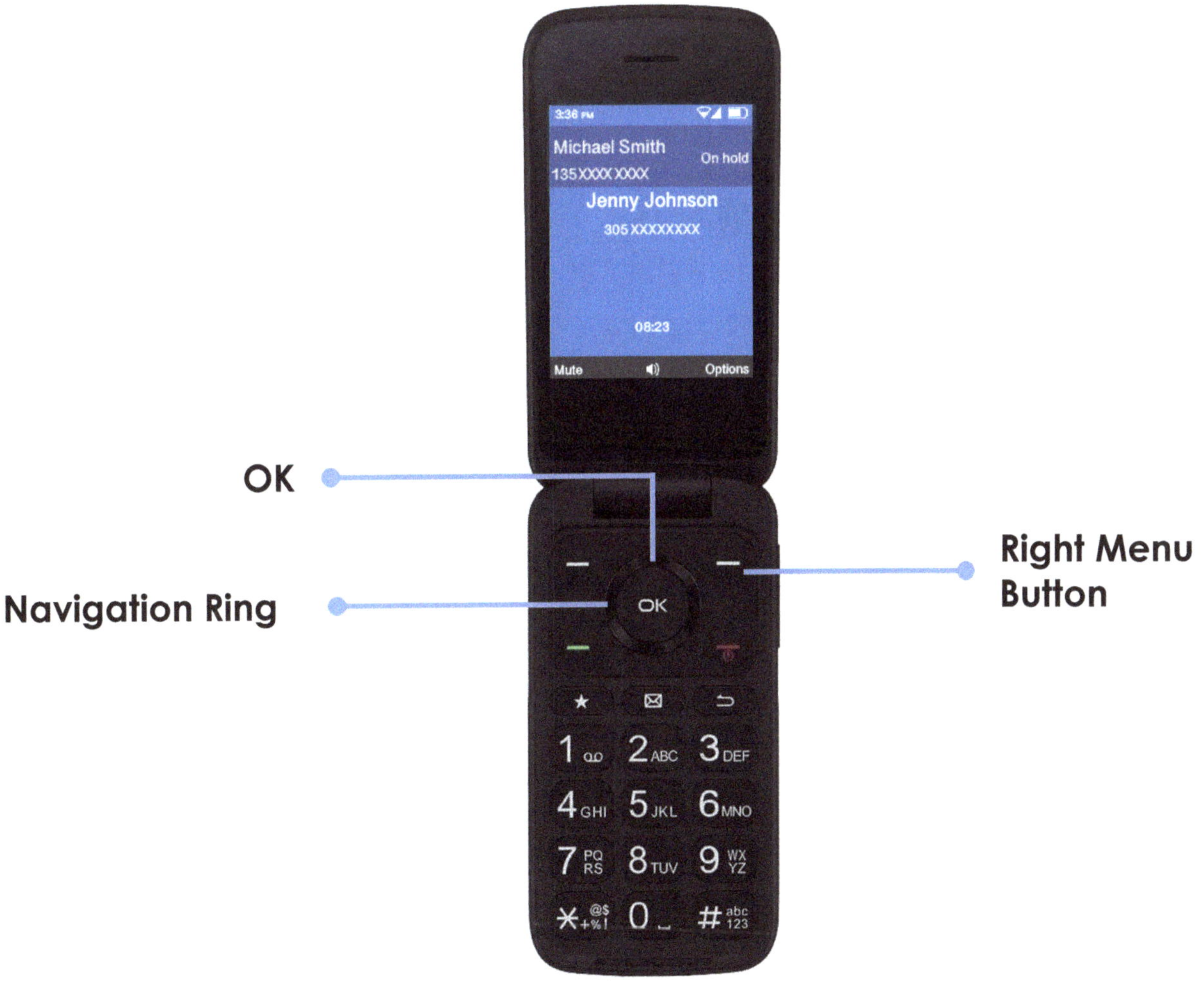

Connecting to a Wi-Fi Network – TCL Phones

8 Joining a Wi-Fi Network

- Connecting to a Wi-Fi network can decrease your usage of cellular data. Note that you may need to enter a password to connect to a Wi-Fi network.

- Open the **Settings**  **Screen** by pressing the **OK Button** from the Home screen.

- Select **Wireless & networks** using the **Navigation Ring**, press **OK** and navigate to **Wi-Fi**, then press **OK**.

- Navigate to **On**, then press **OK** and navigate to **Available networks**, press **OK** and select the **Wi-Fi network** to connect to.

Settings Screen

OK

Navigation Ring

TCL Flip Phone

Answering Incoming Calls – iPhones

1 ### Recognizing an Incoming Call

- When someone calls you, your iPhone will ring, and you'll see the **Caller's Name** or **Phone Number** displayed on your screen.
- You may also hear a ringing or vibrating sound.

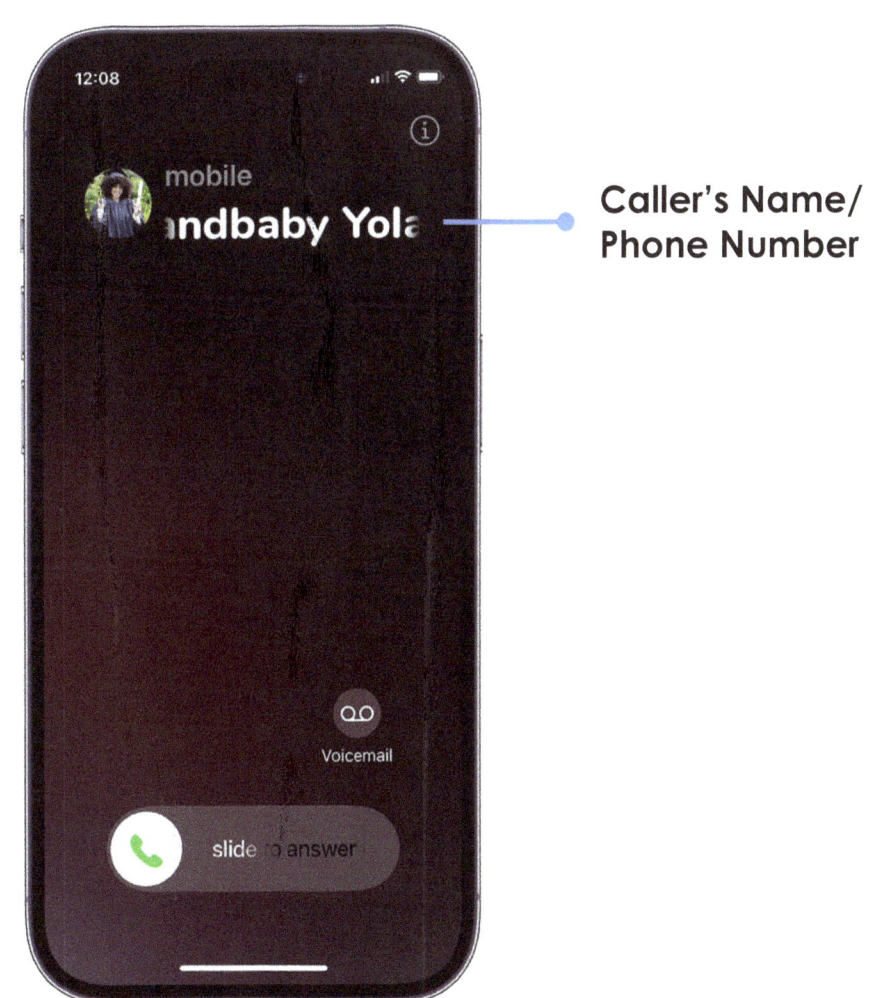

Caller's Name/ Phone Number

② Answering a Call

- To answer the call, slide the **green Answer Button** on the screen to the right. This will connect you to the caller.

- Bring your smartphone close to your ear or use a connected headset to speak to the caller.

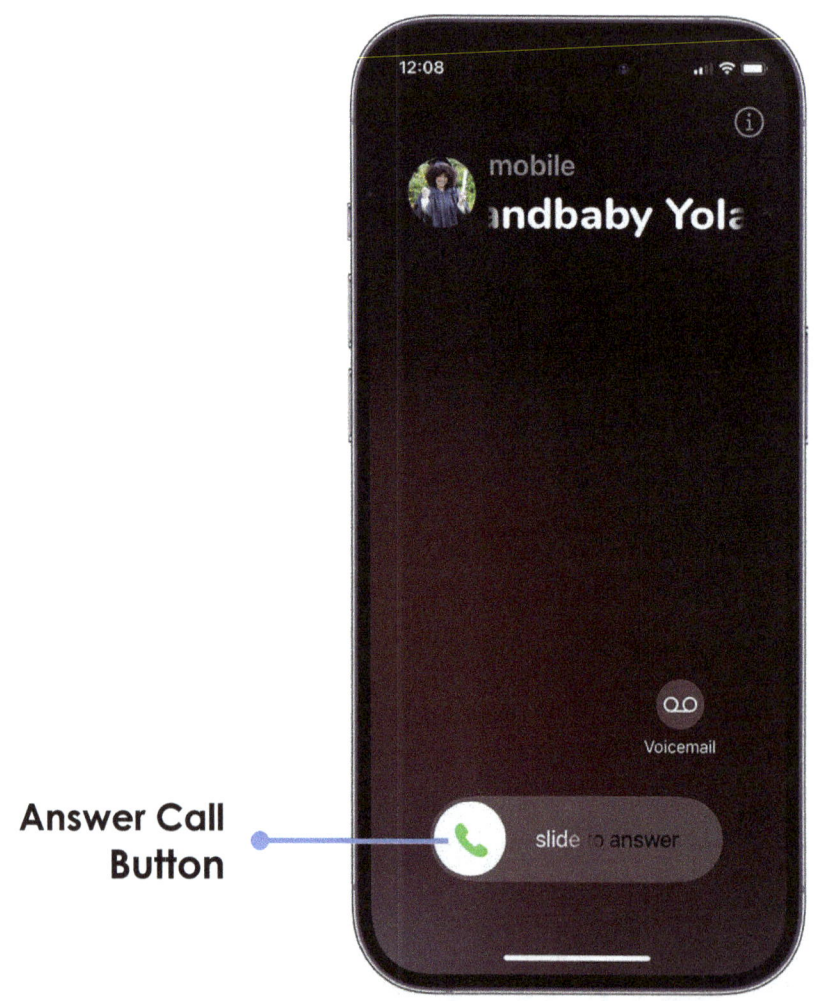

Answer Call Button

3 Ending a Call

- When you're finished with the conversation, tap the **red End Call Button** on the screen to hang up.

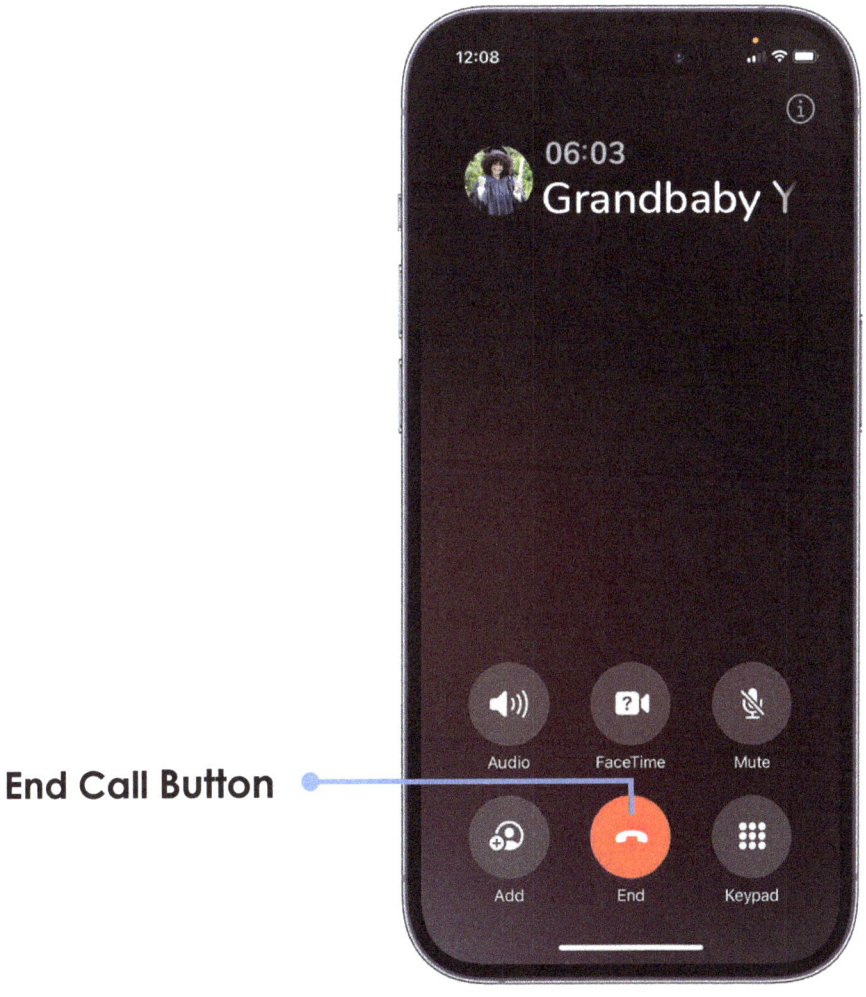

End Call Button

Rejecting Calls – iPhones

Recognizing an Incoming Call You Want to Reject

- If you receive a call from a number you don't recognize or wish to reject, you can decline the call.

- When the call comes in and iPhone is locked, press the **Power/Side Button** twice quickly to reject the call.

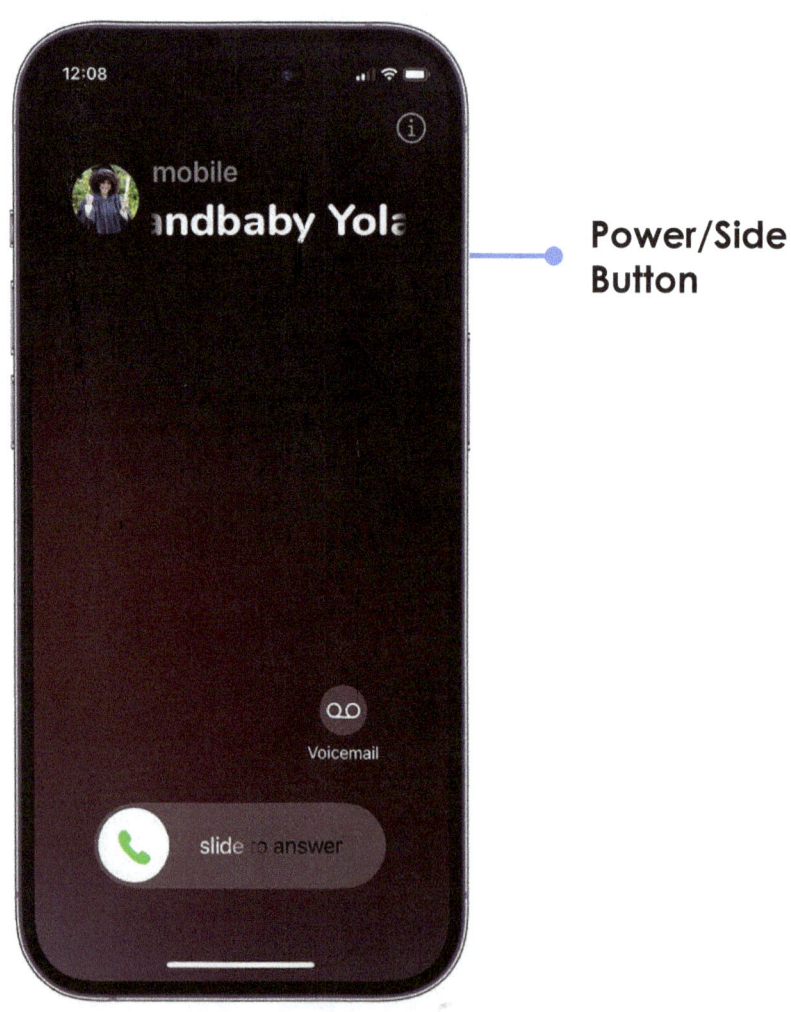

Power/Side Button

 **Rejecting a Call**

- To reject the call when your iPhone is unlocked, tap the **red Decline Call Button** on the screen or swipe up on the **Call Banner**.
- The caller will hear a busy tone and be sent to your voicemail.

Call Banner

Decline Call Button

iPhone

Using Call-Waiting Features – iPhones

 **Managing Call-Waiting**

- When you're on a call and a new call comes in, you'll hear a call-waiting tone.
- You can choose to:
    - End the current call and answer the new one.
    - Put the current call on hold and answer the new call.
    - Ignore the new call and let it go to voicemail.

End Current Call & Answer New Call

Ignore & Send to Voicemail

Put Current Call on Hold & Answer New Call

 **Switching Between Calls**

- If you're on a call and want to switch between active calls (e.g., to alternate between callers), use the **Swap** or **Switch** option on your screen.

- When you hit the **Swap** or **Switch** option, it puts the other caller on hold until you swap back to the original caller.

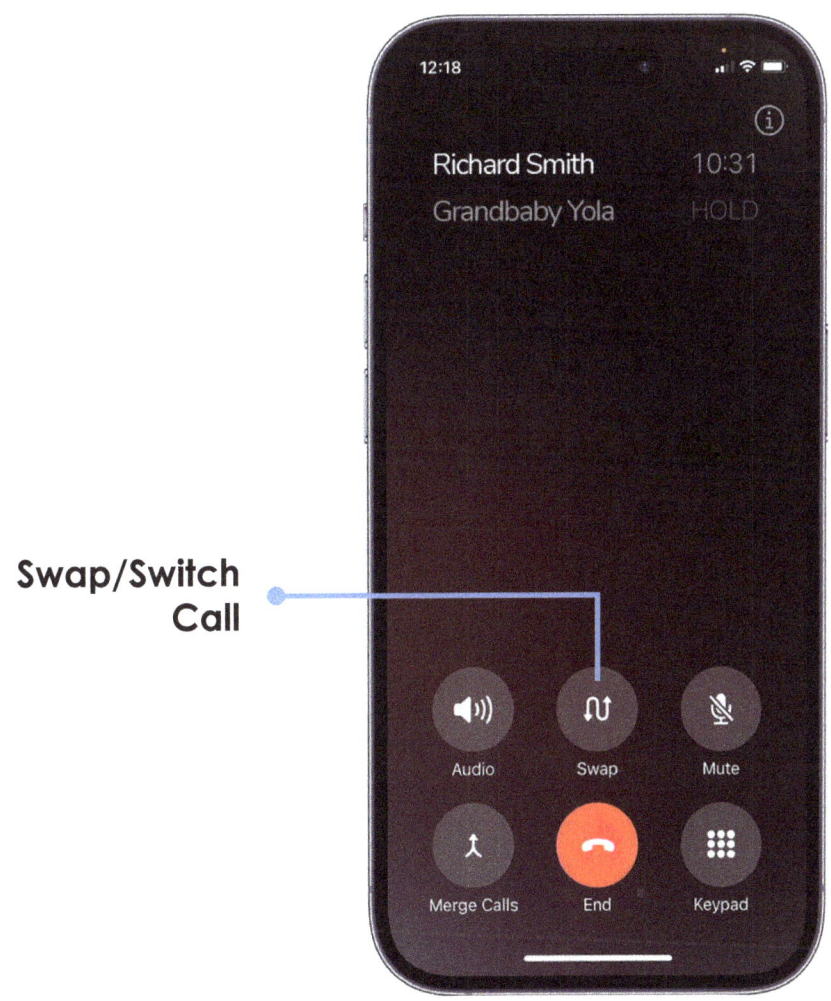

Swap/Switch Call

Connecting to a Wi-Fi Network – iPhones

8 Joining a Wi-Fi Network

- From the **Home Screen**, press the **Settings Icon** .
- Scroll down and select the **Wi-Fi Icon**, then turn on Wi-Fi.
- Tap one of the following:
    - Select a network to join and enter the password, if required.
    - Select Other to join a hidden network, enter the name of the network, security type, password, then press **Join**.

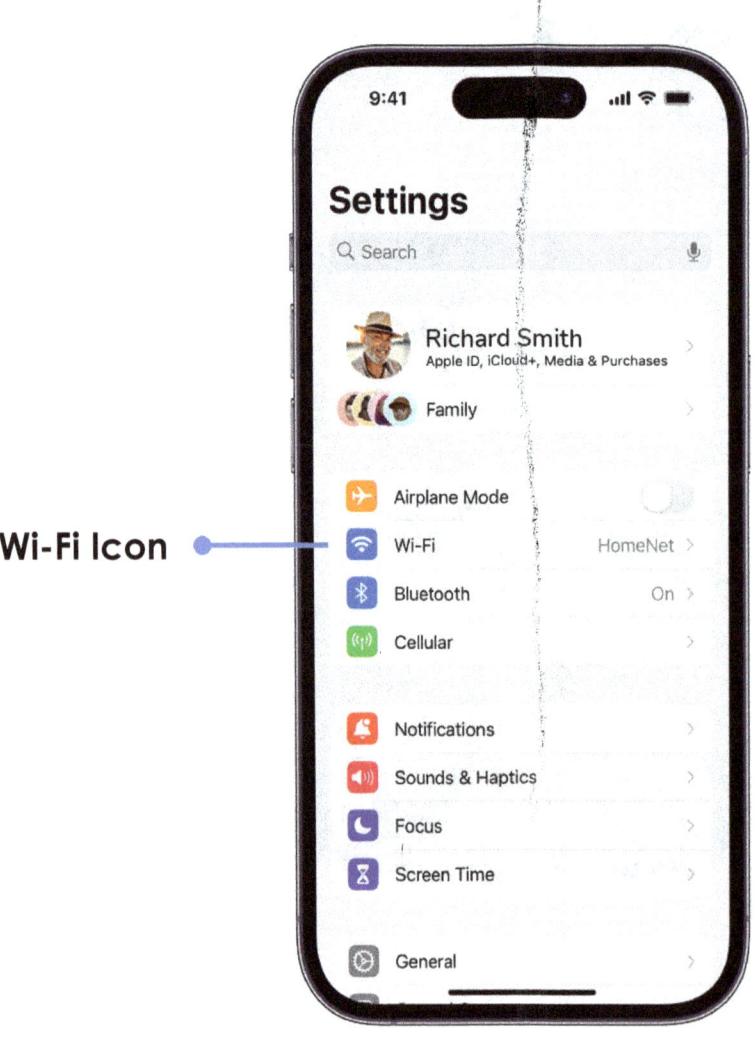

Wi-Fi Icon

Answering Incoming Calls – Galaxy Phones

1 Recognizing an Incoming Call

- When someone calls you, your smartphone will ring, and you'll see the **Caller's Name** or **Phone Number** displayed on your screen.

- You may also hear a ringing or vibrating sound.

Caller's Name/Phone Number

② Answering a Call

- To answer the call, tap the **green Answer Call Button** on the screen. This will connect you to the caller.

- Bring your smartphone close to your ear or use a connected headset to speak to the caller.

Answer Call Button

3 **Ending a Call**

- When you're finished with the conversation, tap the **red End Call Button** on the screen to hang up.

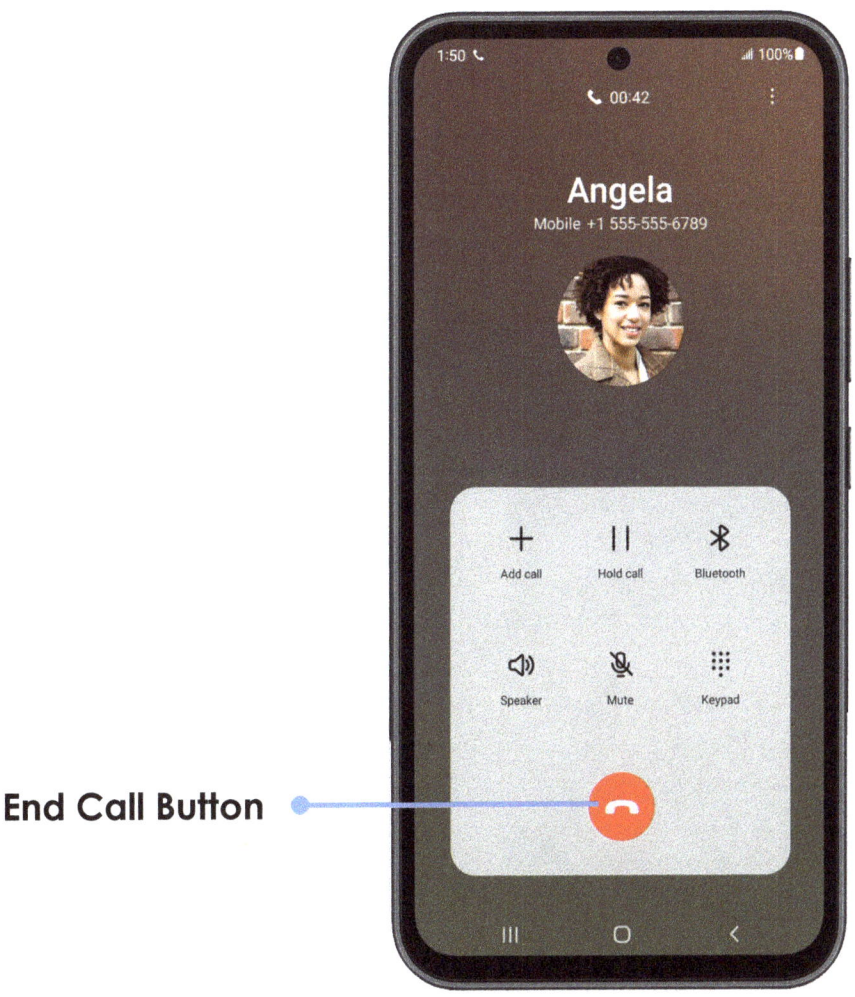

End Call Button

Rejecting Incoming Calls – Galaxy Phones

 **Recognizing an Incoming Call You Want to Reject**

- If you receive a call from a number you don't recognize or wish to reject, you can decline the call.

- When the call comes in, press the **Side Button** twice quickly or tap the **red Decline Button**, or swipe up on the **t** at the top of your screen.

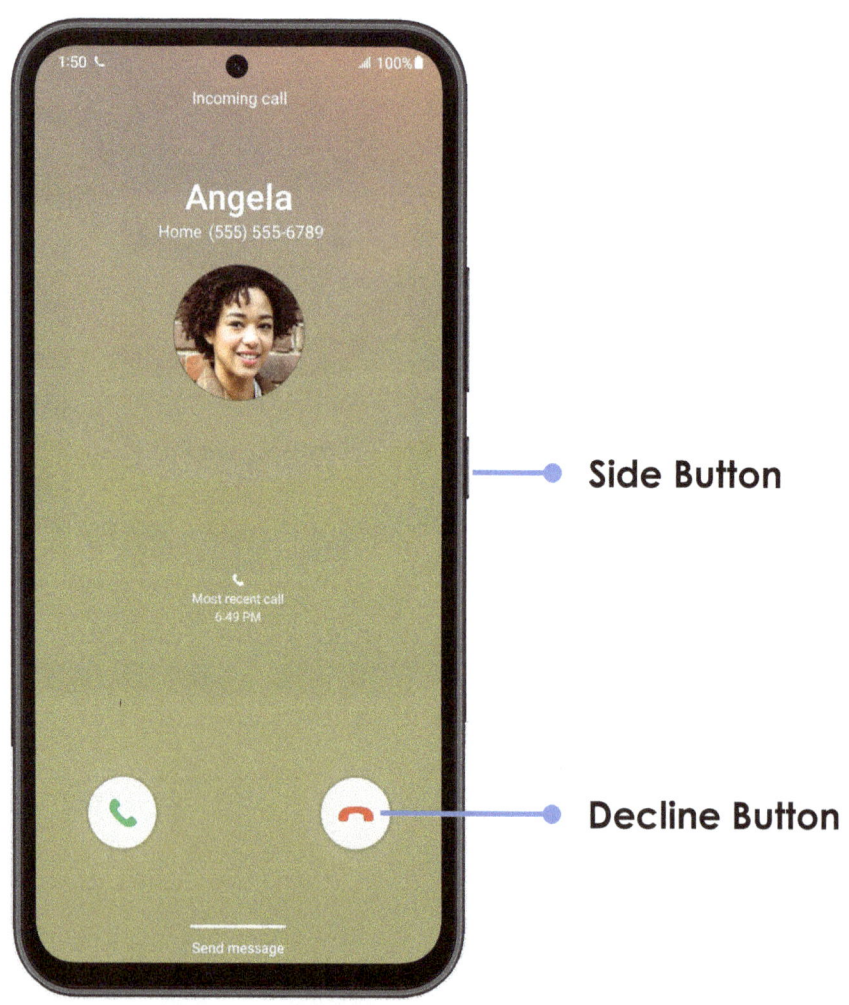

 Screening Calls

- If you are unable to take a phone call, you can ignore the phone call and reply with a text to let the caller know you are unable to talk at that time.

- From an incoming call, drag up the **Send Message Bar**.

- Select one of the preprogrammed responses or Tap **Write a New Message**.

- Enter your message and tap the **Send Icon** when finished.

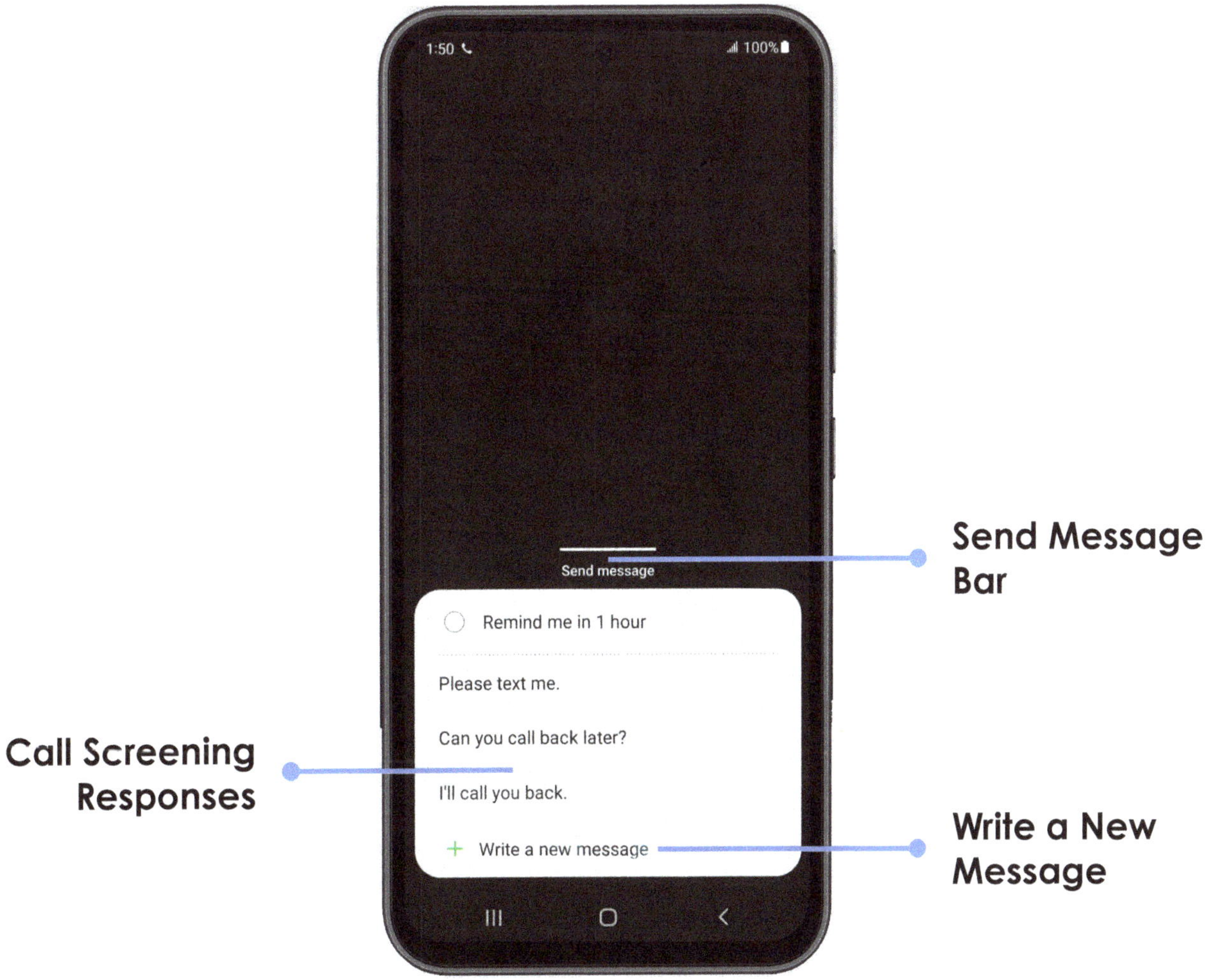

Send Message Bar

Call Screening Responses

Write a New Message

Using Call-Waiting Features – Galaxy Phones

 **Managing Call-Waiting**

- Call-waiting allows you to receive a new call while you're already on one.

- During an ongoing call, drag the **green Answer Call Button** up to answer the incoming call. (Note: This will put the original call on hold.)

- Select the call on hold to return to it.

- Tap the **End Call Button** to end the current call. (Note: When you end one call, you'll automatically switch to the other call.)

Answer New Call

End Call Icon

7 Switching Between Calls

- If you're on a call and want to switch between active calls (e.g., to alternate between callers), tap the **Swap Icon**. (Note: This will put the original call on hold.)

- Tap the **End Call Icon** to end the current call. (Note: When you end one call, you'll automatically switch to the other call.)

Connecting to a Wi-Fi Network – Galaxy Phones

Joining a Wi-Fi Network

- From the **Home Screen**, drag down the **Notification Panel**.
- Touch and hold the **Wi-Fi Icon**, then press the Wi-Fi switch to **ON**.
- Select the Wi-Fi Network you want to connect to and enter the password, if required, then press **Connect**.

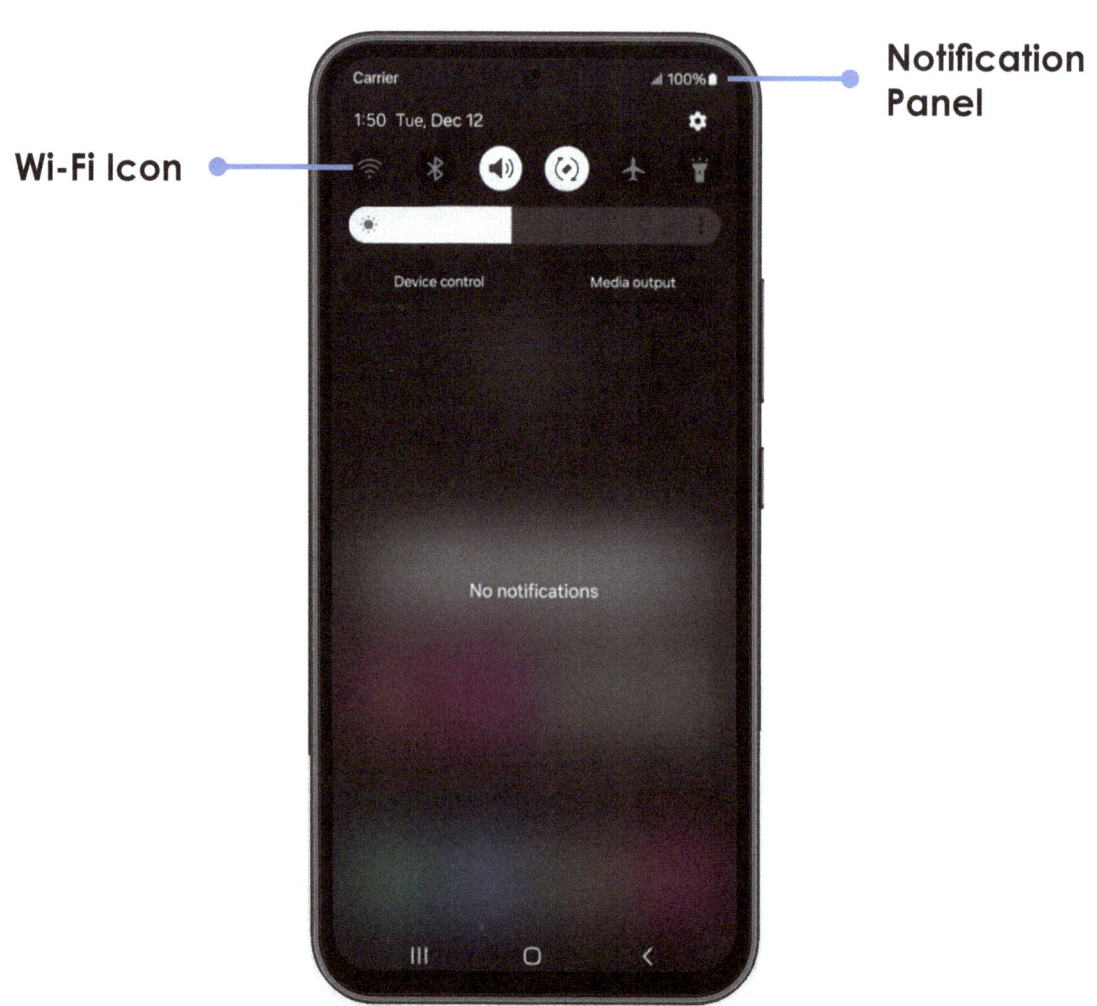

Chapter Questions

1. When you are receiving an incoming call, your phone will ring or vibrate.

    ○ True
    ○ False

2. To answer an incoming call on a smartphone, what do you do?

    ○ Ignore the call
    ○ Tap the green "answer" button on the screen
    ○ Tap the decline button

3. To end or finish a phone conversation on a smartphone, tap the accept button to hang up.

    ○ True
    ○ False

ANSWERS: 1. TRUE | 2. Tap the green "answer" button on the screen | 3. FALSE - Tap the end button to hang up

CHAPTER 3

Sending Text Messages

Introducing Seniors to Sending and Managing Text Messages

Composing and Sending Text Messages: Hands-On Practice

Composing a Text Message – TCL Phones

In this hands-on training, we'll walk you through the process of composing and sending text messages on your basic phone. Text messaging is a convenient way to stay in touch with loved ones and friends. By the end of this guide, you'll be more comfortable and confident in sending and receiving text messages.

Unlocking Your Basic Phone

- **Open the Flip** to wake up your basic phone or press the **Power/Hang Up Button** to turn on your basic phone if off.

- If your phone has a security feature like a PIN, unlock it by using the **Keypad**.

Flip Open/Close

Power/Hang Up Button

Keypad

TCL Flip Phone

89

② Accessing the Messages App

- Press the **Text Message Button** on the **Keypad** or from the **Home Screen**, press **OK** and select **Messages App** .

Home Screen

OK

Text Message Button

Keypad

TCL Flip Phone

3 Starting a New Message

- Inside the **Messages App** , press the **Left Menu Button** to create a **New Message**.

Left Menu Button

Selecting a Contact

- Use the **Navigation Ring**, move up or down to highlight the **To Field**, then use the **Keypad** to enter the phone number of the recipient in the **"To" Field**, or press **Right Menu Button** to access **Options**.

- Highlight **Select from Contact** or **Select Contact Group**, press **OK** then navigate to the desired contact to add a contact from the contacts list.

Composing a Message

- Using the **Navigation Ring**, move down to access the **Message Field**, then use the **Keypad** to enter the desired message.

Message Field

Navigation Ring

Keypad

TCL Flip Phone

93

Sending Your Message

- When you finish composing a message, press **OK** to select **Send** or the **Right Menu Button** to select **Options**, then **Send** to send the message.

- To add additional recipient(s) and create a **Group Message**, enter the additional contact(s), then use the **Navigation Ring** to move down to highlight the **Message Field**.

- To remove recipient(s), press the **Back/Delete Key** to delete the desired contact(s).

Message Field

OK

Navigation Ring

Right Menu Button

Back/Delete Key

Receiving and Responding to Text Messages – TCL Phones

Receiving a Text Message

- When someone sends you a text message, your phone will notify you with a sound, vibration, or notification banner.

- To read the message, press the **Left Menu Button** to view **Notifications**. Use the **Navigation Ring** to highlight the message thread and press **OK** to open.

Responding to a Text Message

- From the desired **Message**, use the **Navigation Ring** to highlight the **Message Field**, then use the **Keypad** to enter the desired response in the **Message Field**. When finished, press **OK** to select **Send**.

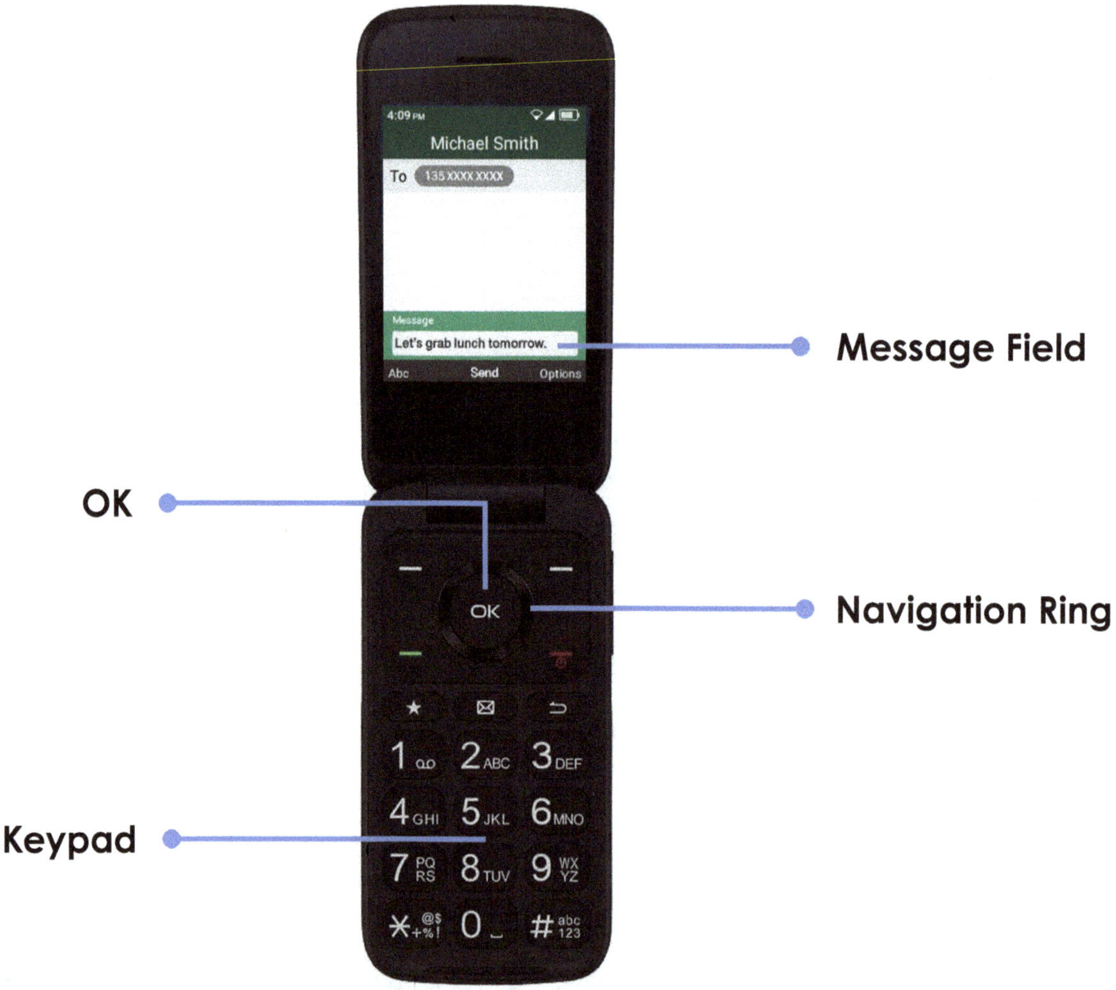

Message Field

OK

Navigation Ring

Keypad

Adding Attachments (Optional)

- If you want to send pictures, videos, or other files, highlight the **Message Field**, press the **Right Menu Button** to select **Options**.

- Use the **Navigation ring** to highlight **Add attachment**.

- Use the **Navigation ring** to find the attachment, then press **OK**.

Message Field

OK

Navigation Ring

Right Menu Button

Composing a Text Message – iPhones

In this hands-on training, we'll walk you through the process of composing and sending text messages on your smartphone. Text messaging is a convenient way to stay in touch with loved ones and friends. By the end of this guide, you'll be more comfortable and confident in sending and receiving text messages.

 **Unlocking Your Smartphone**

- Press the **Power/Side Button** on your smartphone to wake it up.
- If your phone has a security feature like a **PIN**, **Pattern ID**, **Touch ID** (fingerprint), or **Face ID** lock, unlock it by following the on-screen prompts.

Power/Side Button

2 **Accessing the Messages App**

- Look for the **Message App Icon** 💬 on your home screen or in your **App Drawer**. It often looks like a speech bubble or a messaging symbol.

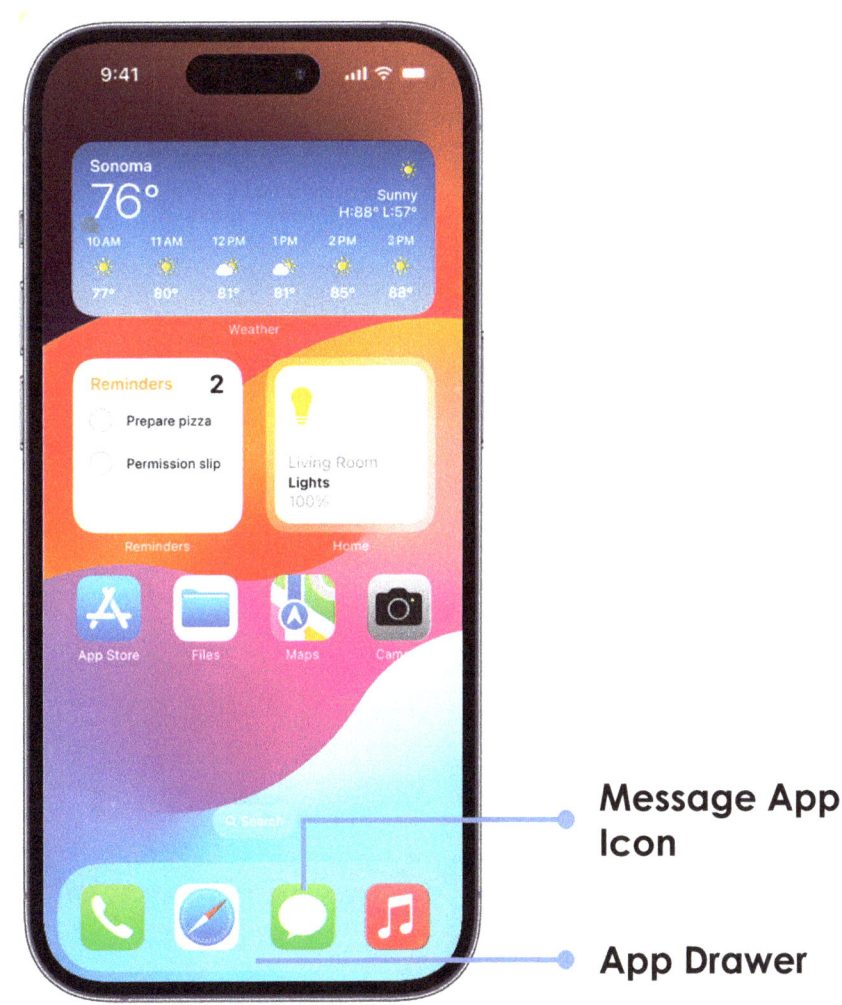

3 Starting a New Message

- Inside the **Message App** , locate the **Compose** or **New Message Icon**, usually represented by a **Pencil** (in a box) or a "**+**" sign. Tap it to start a new message.

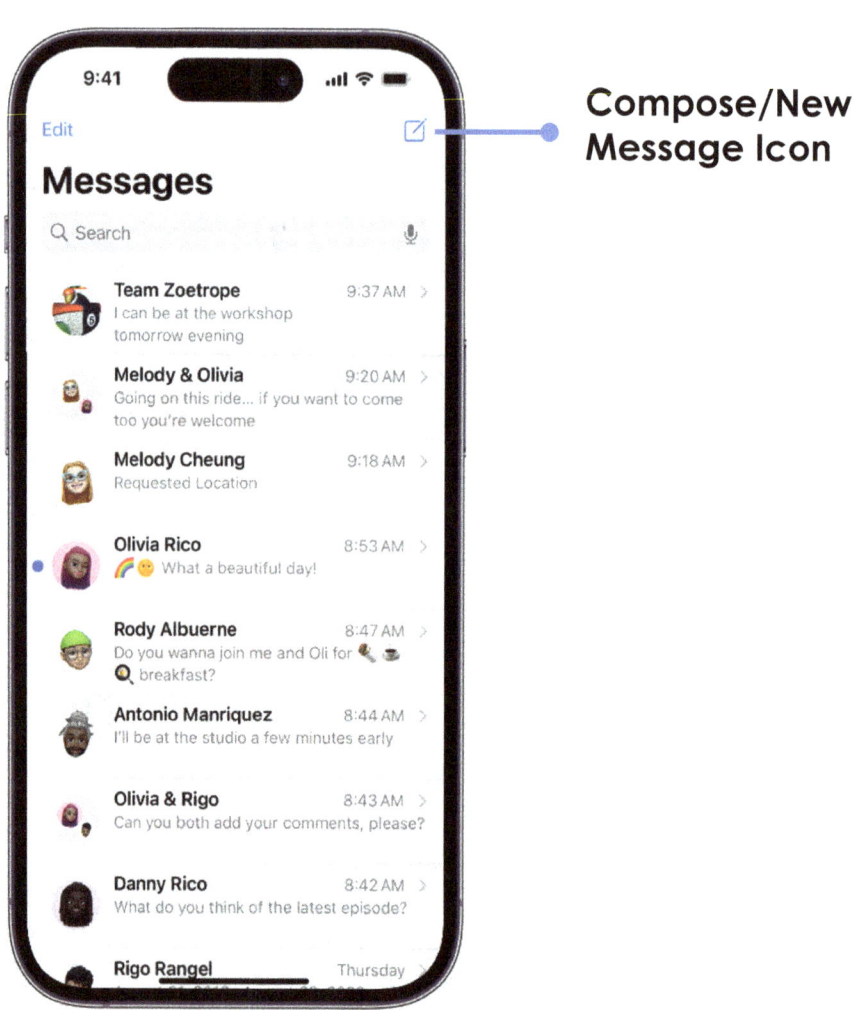

Compose/New Message Icon

iPhone

 **Selecting a Contact**

- To send a message to someone in your contacts, tap the **To Field**.

- A list of your contacts will appear. Scroll through the list or use the **Search Bar** to type the contact's name to find the information. Next, tap the person's name to select the contact.

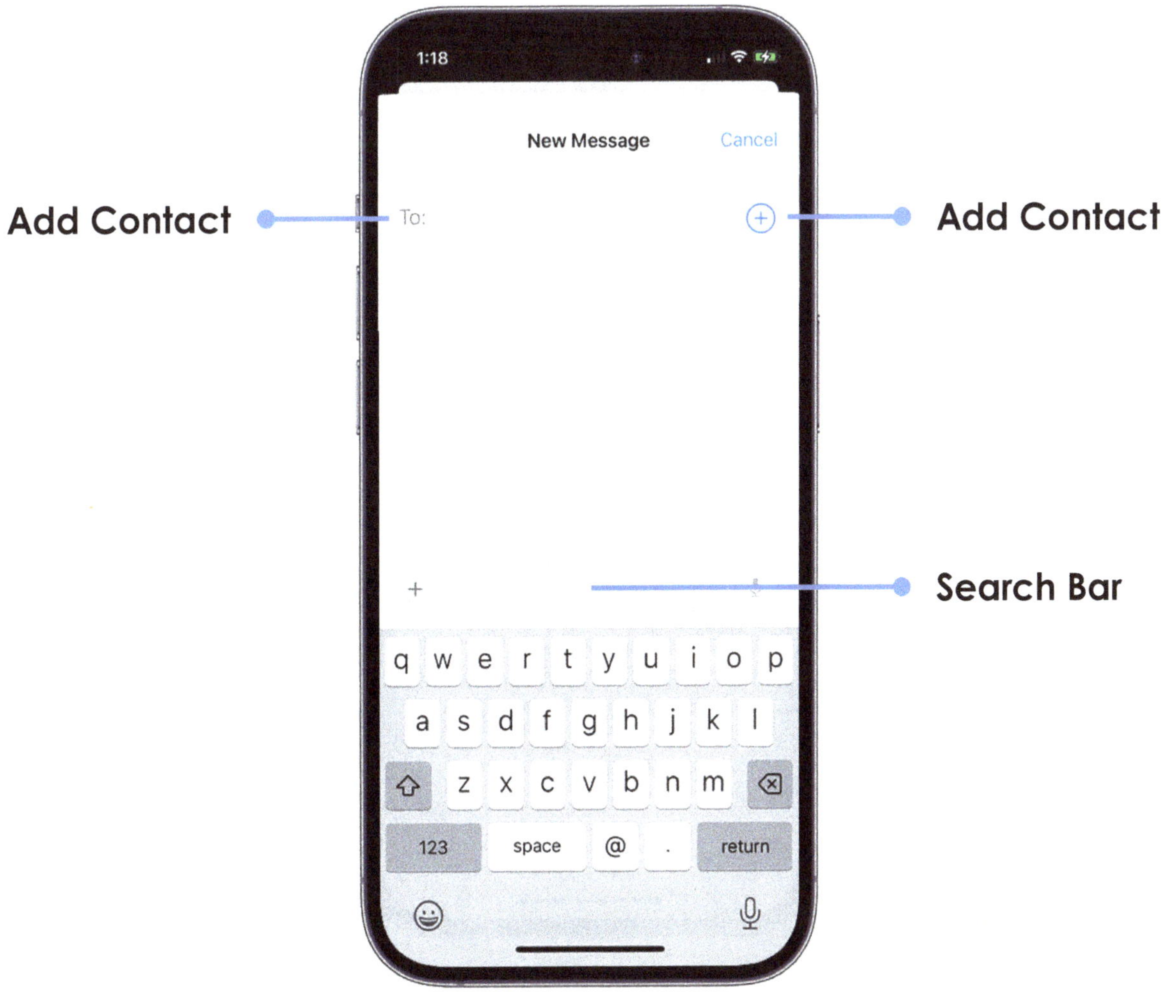

Add Contact — To:

Add Contact — ⊕

Search Bar

iPhone

Composing Your Message

- Tap the **Text Field** at the bottom of the screen to start typing your message.

- Use the **Keyboard** to enter your message. You can type using the on-screen keyboard, which pops up when you tap the text field.

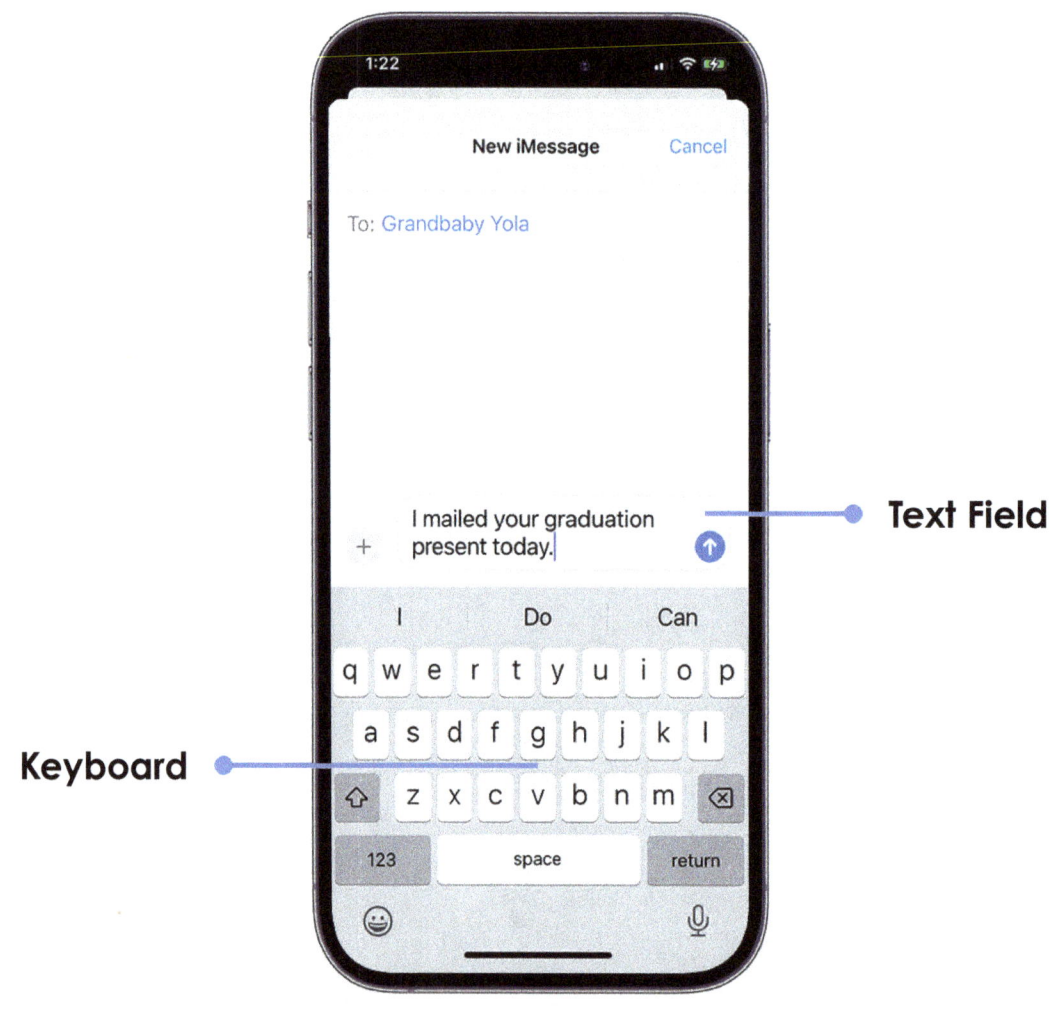

 **Sending Your Message**

- Once your message is ready, tap the **Send Button** (usually represented by a paper airplane icon, up arrow, or "Send").

- Your message will be sent to the recipient.

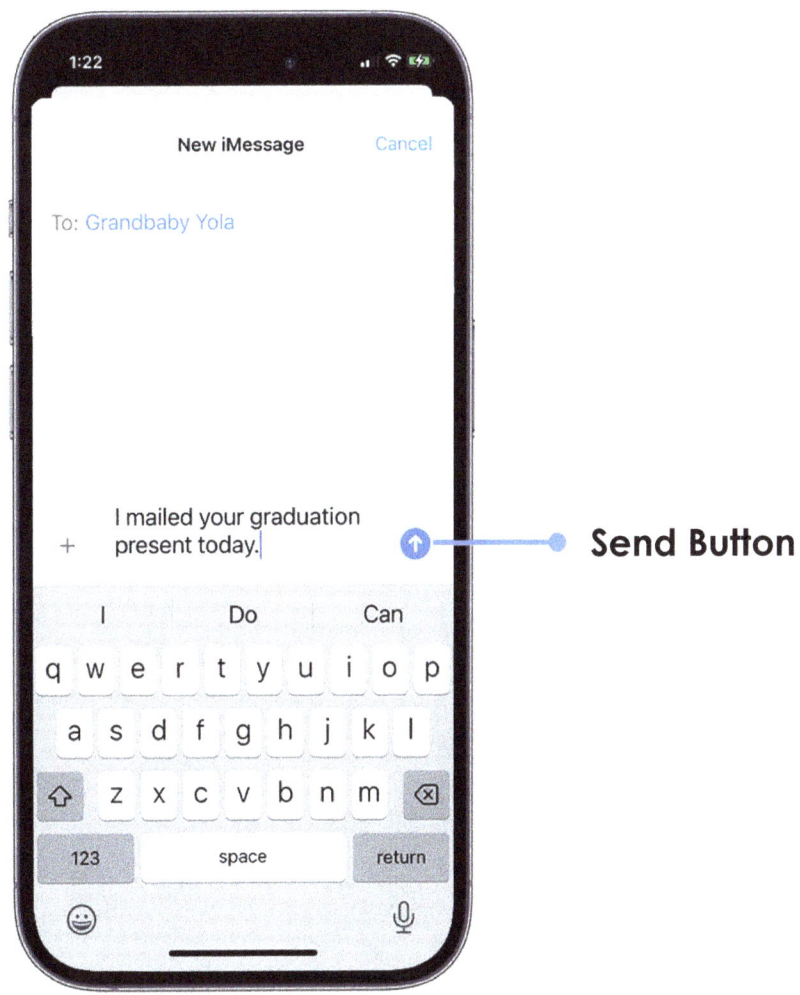

Send Button

Receiving and Responding to Text Messages – iPhones

7 Receiving a Text Message

- When someone sends you a text message, your phone will notify you with a sound, vibration, or notification banner.

- To read the message, simply tap the notification or open your **Message App Icon**  and look for the blue dot that indicates an **Unread Message**.

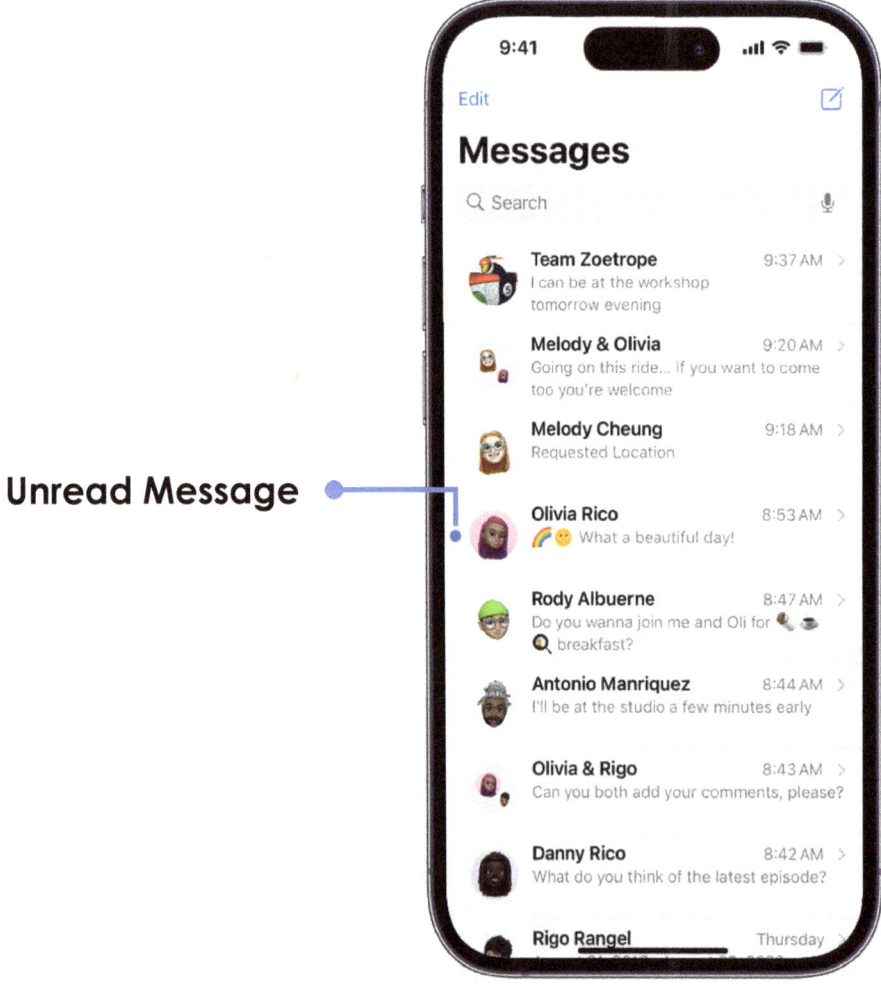

Unread Message

 **Responding to a Text Message**

- Open the conversation with the person who sent you a message.
- Tap the **Text Field** at the bottom and type your response.
- Tap the **Send Button** to send your reply.

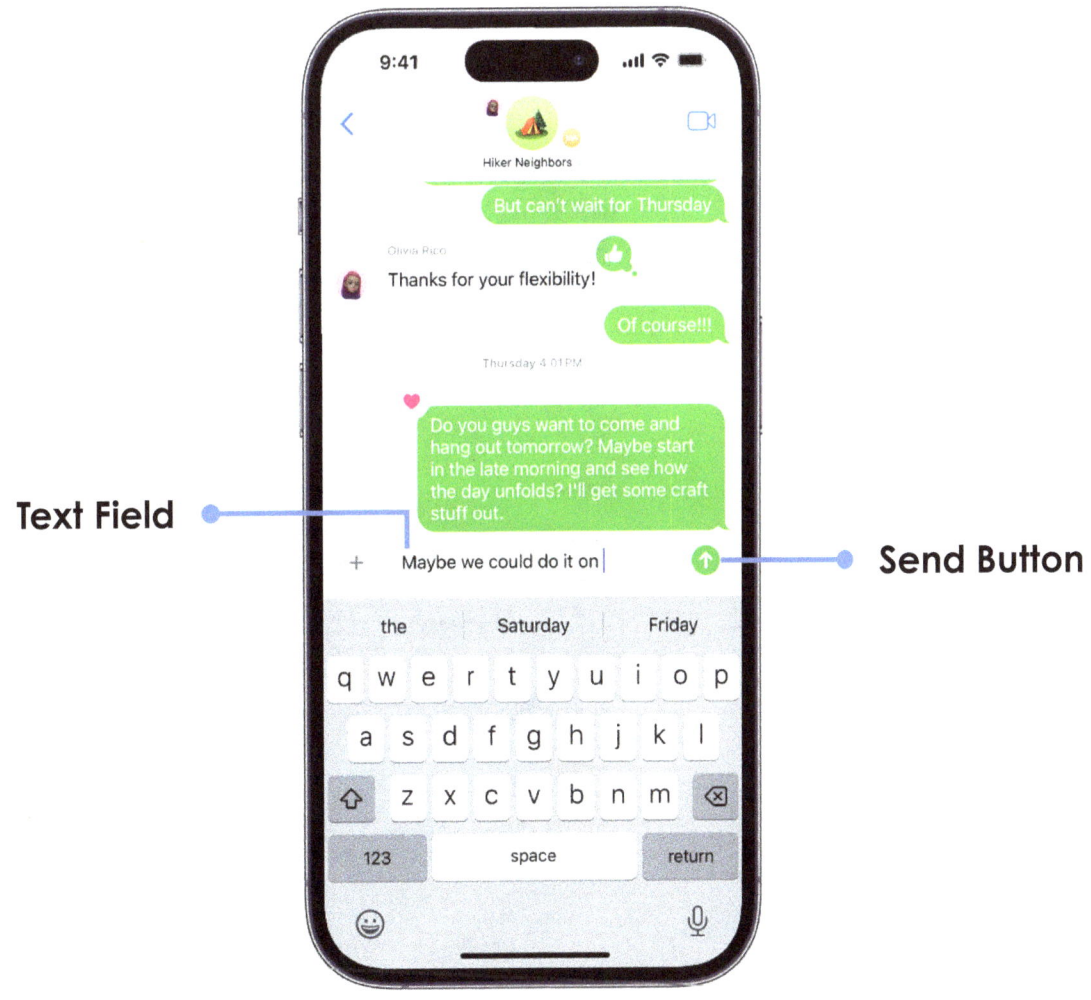

9 Adding Attachments (Optional)

- If you want to send pictures, videos, or other files, look for an **Attachment Icon** (usually a "+" sign, paperclip, or camera icon) in the **Message App** .

- Tap the **Attachment Icon**, select the file you want to send, and then tap **Send**.

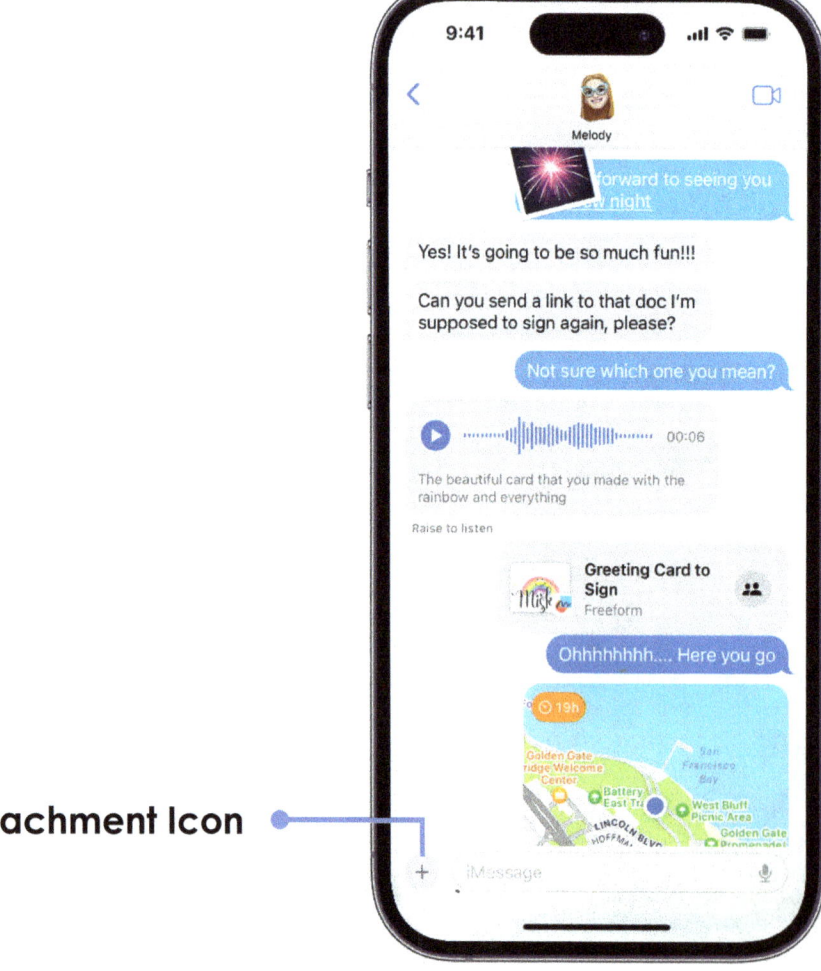

Attachment Icon

Composing a Text Message – Galaxy Phones

In this hands-on training, we'll walk you through the process of composing and sending text messages on your smartphone. Text messaging is a convenient way to stay in touch with loved ones and friends. By the end of this guide, you'll be more comfortable and confident in sending and receiving text messages.

1 Unlocking Your Smartphone

- Press the **Power/Side Button** on your smartphone to wake it up.
- If your phone has a security feature like a **PIN**, **Pattern ID**, **Touch ID** (fingerprint), or **Face ID** lock, unlock it by following the on-screen prompts.

Power/Side Button

② Accessing the Message App

- Look for the **Message App Icon** on your home screen or in your **App Drawer**. It often looks like a speech bubble or a messaging symbol.

 **Starting a New Message**

- Inside the **Message App** 💬, locate the **Start Chat Button** and tap it to start a new message.

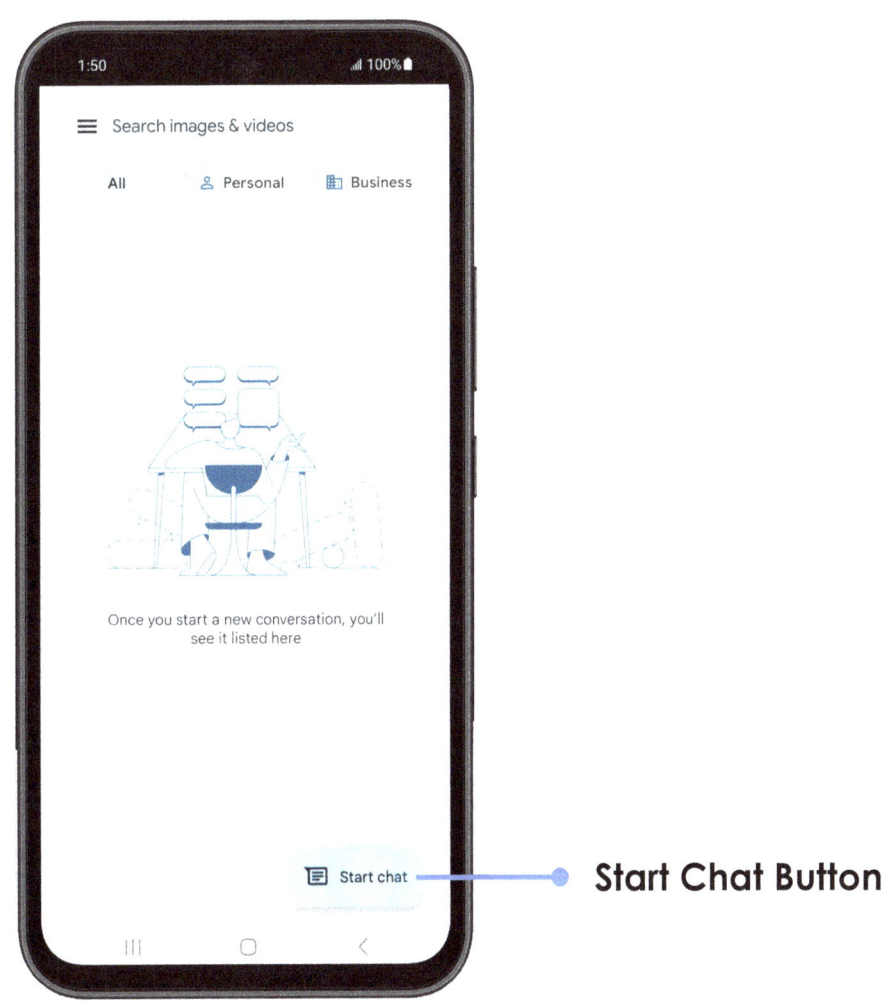

Start Chat Button

Selecting a Contact

- To send a message to someone in your contacts, tap the **To Field**.

- Scroll through the list or use the search bar (by typing in the person's name) to find the person you want to message.

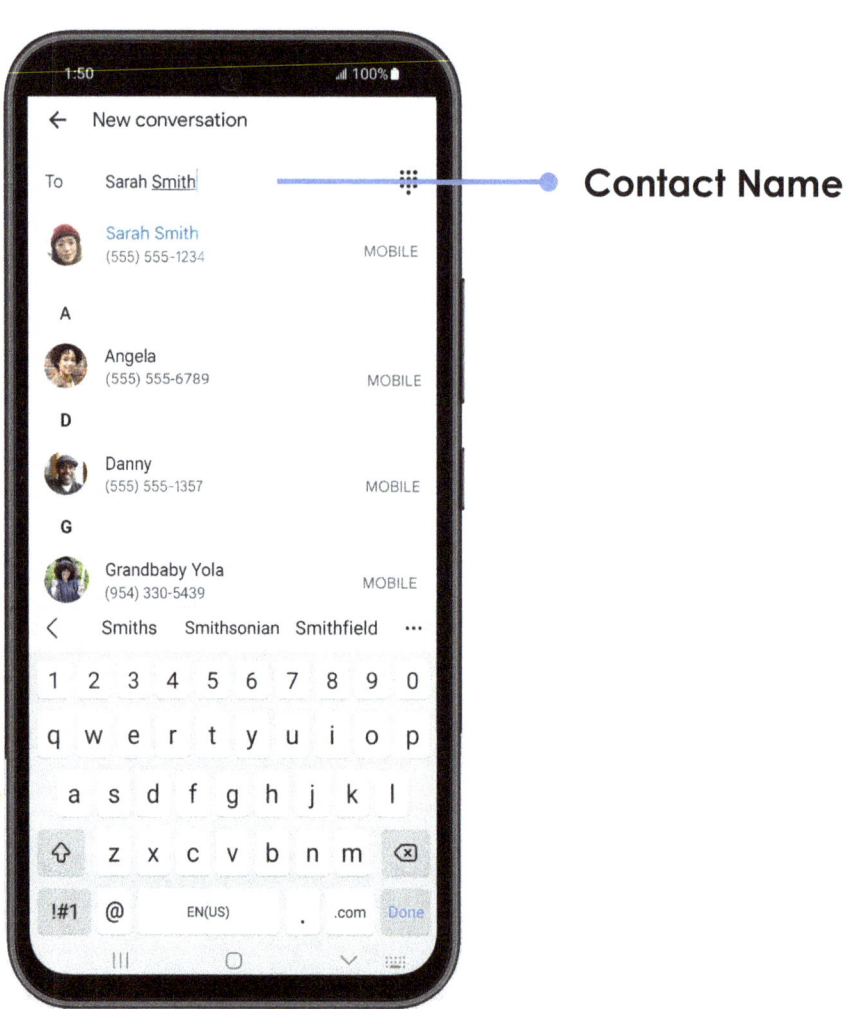

Contact Name

Composing Your Message

- Tap the **Text field** at the bottom of the screen to start typing your message.

- Use the **Keyboard** to enter your message. You can type using the on-screen keyboard, which pops up when you tap the text field.

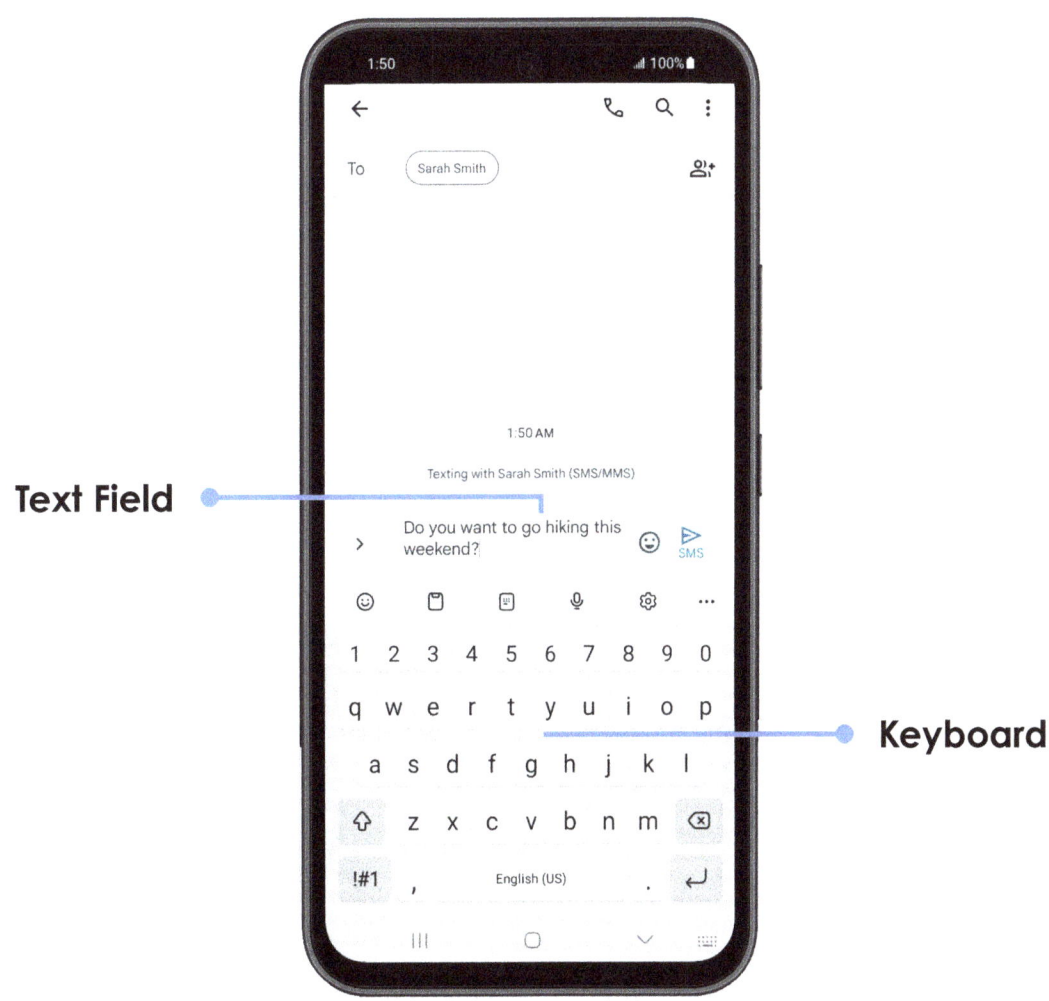

 **Sending Your Message**

- Once your message is ready, tap the **Send Text Message Icon** (usually represented by a paper airplane icon or "Send").

- Your message will be sent to the recipient.

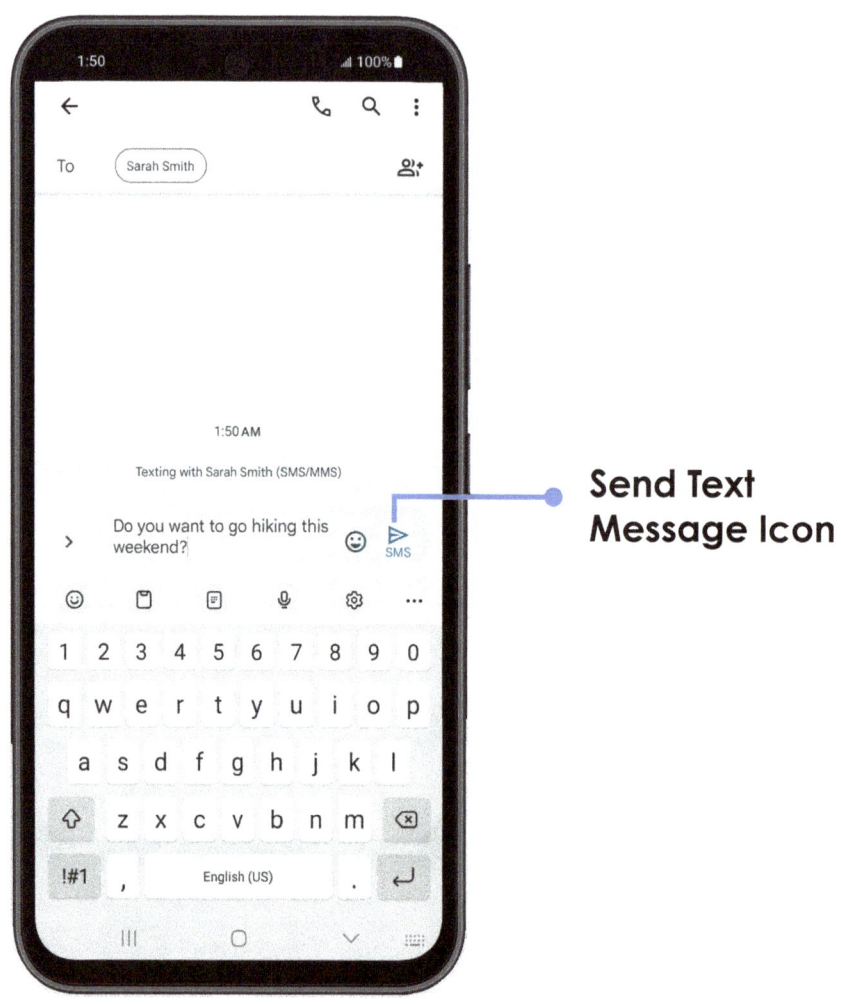

Send Text Message Icon

112

Receiving and Responding to Text Messages – Galaxy Phones

Receiving a Text Message

- When someone sends you a text message, your phone will notify you with a sound, vibration, or notification banner.

- To read the message, simply tap the **Notification Banner** or open your **Message App** 💬 and look for the conversation.

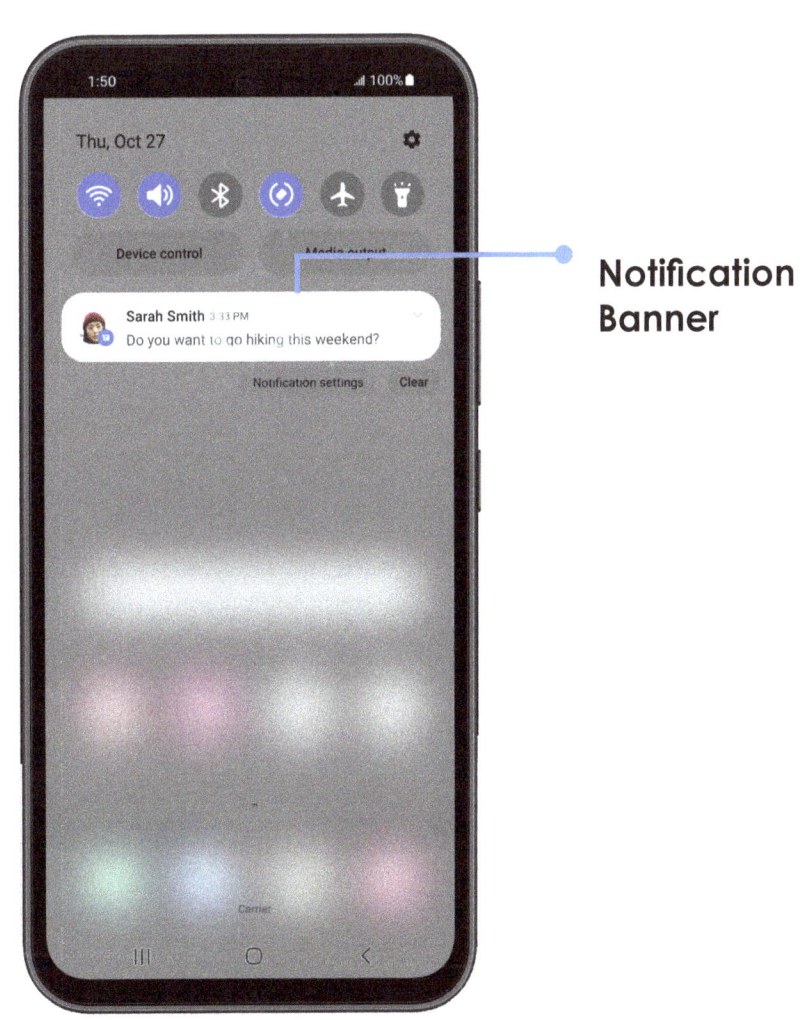

Notification Banner

Responding to a Text Message

- Open the conversation with the person who sent you a message.
- Tap the **Text Field** at the bottom and type your response using the on-screen **Keyboard**.
- Tap the **Send Text Message Icon** to send your reply.

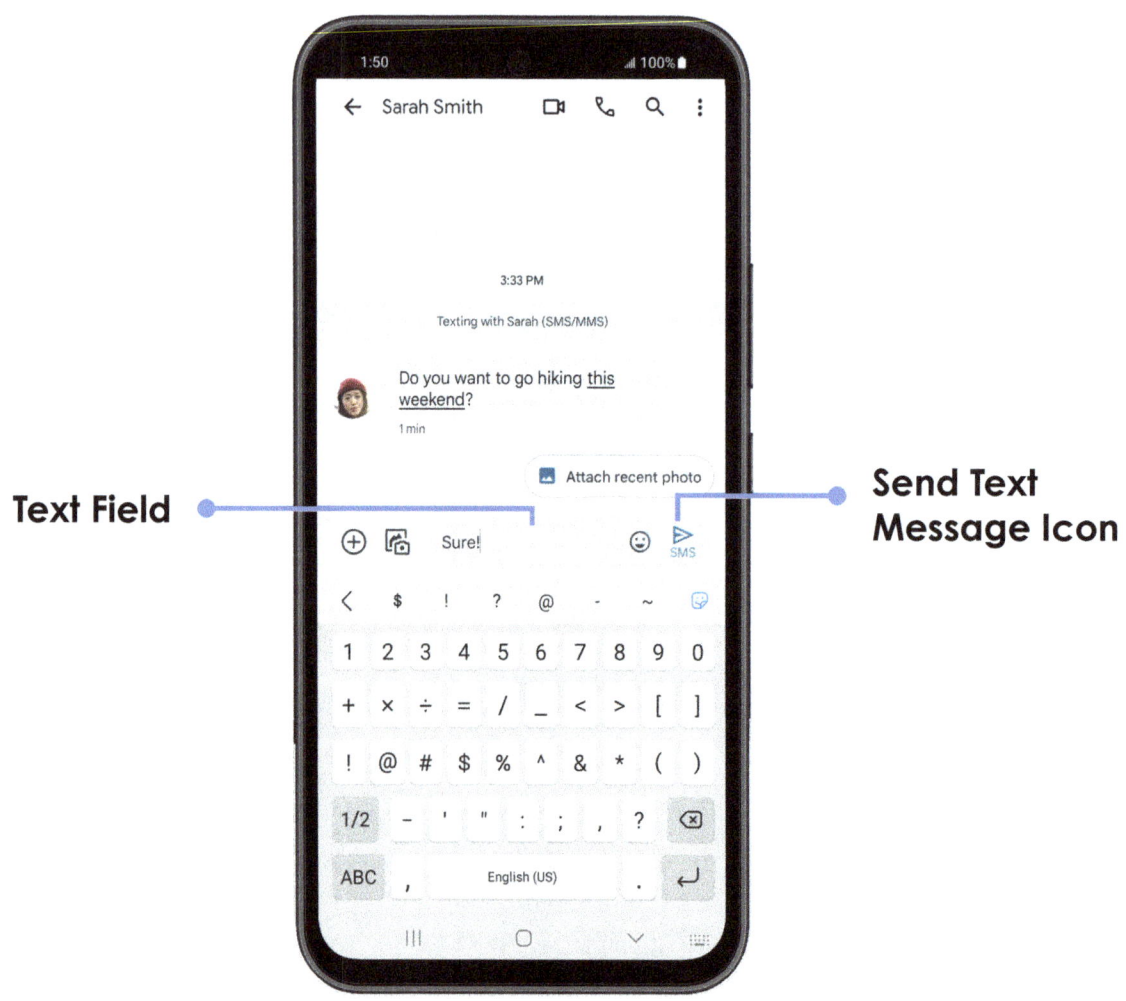

Text Field

Send Text Message Icon

 **Adding Attachments (Optional)**

- If you want to send pictures, videos, or other files, look for an attachment icon (usually a paperclip or camera icon) in the **Message App** 💬.

- Tap the **Attachment Icon**, select the file you want to send, and then tap the **Send Text Message Icon**.

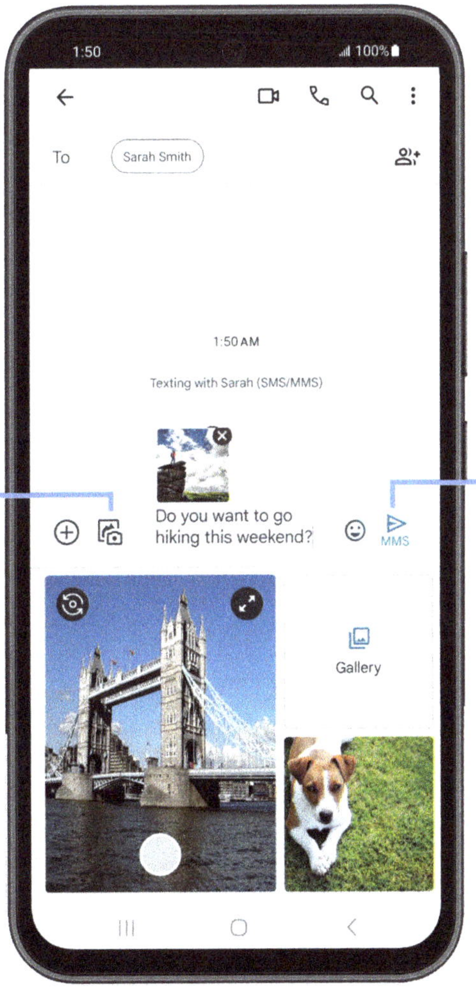

Attachment Icon

Send Text Message Icon

Chapter Questions

Chapter questions apply for any cell phones covered.

1. To send an attachment what steps do you take once you compose the text?

    ○ Attach file and submit

    ○ Submit and attach file

    ○ None of the above

2. To start a new message, go into the contact app first.

    ○ True

    ○ False

3. Once you have typed your message, it will automatically send when you are finished typing.

    ○ True

    ○ False

4. When you receive a text message, tap the notification button or open your message app and look at the conversation.

    ○ True

    ○ False

ANSWERS: 1. Attach file and submit | 2. FALSE - Go in the message app, locate the compose or new message button and tap to start a new message. | 3. FALSE - Once message is ready, tap the send button (usually paper airplane), your message will then be sent to the recipient | 4. TRUE

NOTES

CHAPTER 4
Taking Pictures, Selfies, and Videos

An Introduction to Taking Pictures, Selfies, and Shooting Videos

Taking Pictures, Selfies, and Shooting Videos: Hands-On Practice

Taking Pictures – TCL Phones

This chapter is a step-by-step process for capturing moments with your cell phone! In the era of advanced technology, your mobile device is not just a communication tool but also a powerful photography companion.

 **Unlocking Your Basic Phone**

- **Open the Flip** to wake up your basic phone or press the **Power/Hang Up Button** to turn on your basic phone if off.

- If your phone has a security feature like a **PIN**, unlock it by using the **Keypad**.

Home Screen

Flip Open/Close

Power/Hang Up Button

Keypad

TCL Flip Phone

② Accessing the Camera App

- From the **Home Screen**, press **OK** to go to the **Apps Screen**.
- Use the **Navigation Ring** to select the **Camera App** and press **OK**.

Apps Screen

Camera App

OK

3 Taking a Picture

- Position the object or landscape in the screen, and press **OK** to **Take the Photo**.
- Photos will be automatically save to the **Gallery app**.
- After taking photo, press **Left Menu Button** to preview.

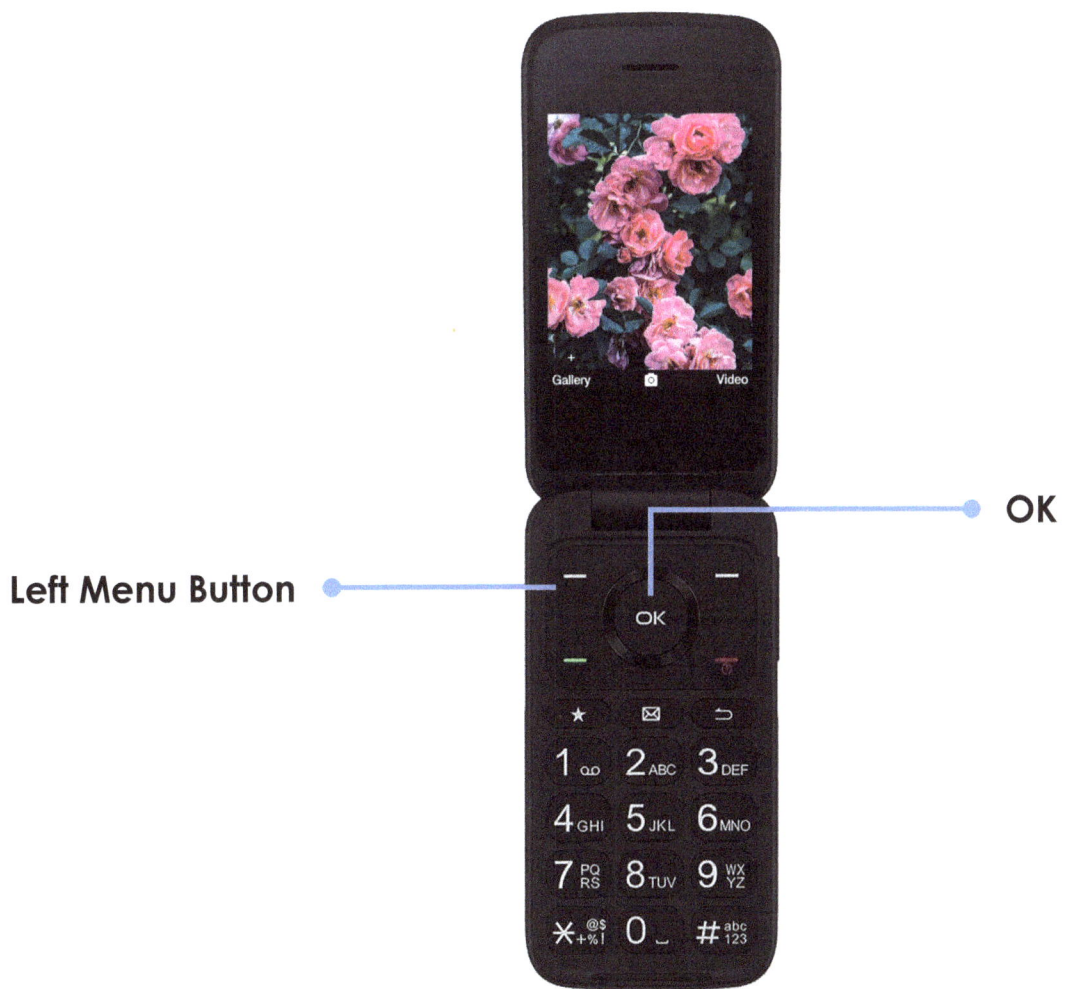

Zooming In & Out

- While positioning the object or landscape in the screen before taking a picture, press **Volume Up/Down Key**, or press "*" and "#" to zoom lens in and out.

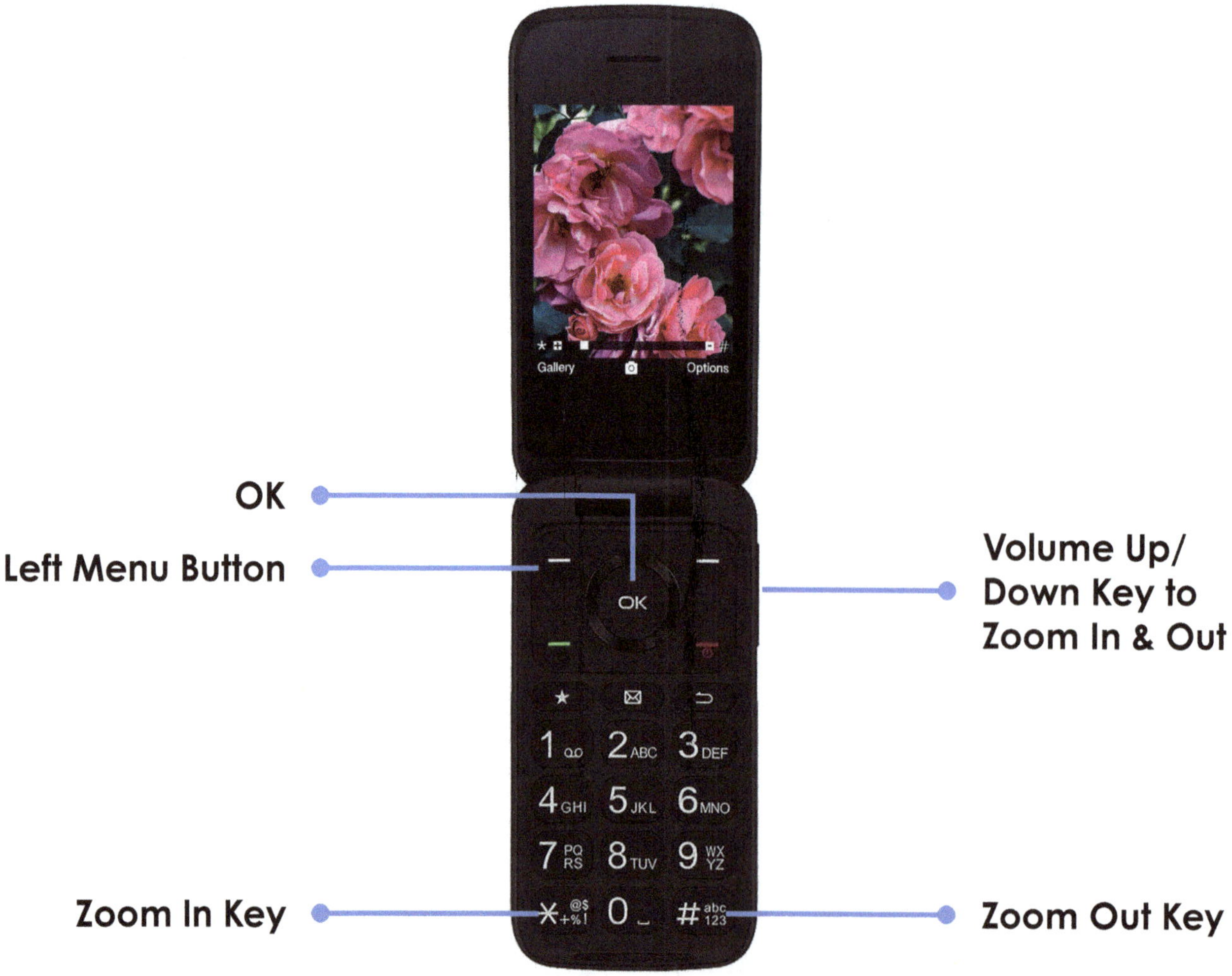

OK

Left Menu Button

Volume Up/Down Key to Zoom In & Out

Zoom In Key

Zoom Out Key

Shooting Videos – TCL Phones

5 Switching to Video Mode

- While in **Photo Mode** from the camera screen, press the **Right Menu Button** to switch to **Video Mode**.

Right Menu Button

Shooting a Video

- Position the object or landscape in the screen, and press **OK** to **Record a Video**.

- Press **Volume Up/Down key**, or press "*" and "#" to zoom lens in and out.

- Press **OK** to **End Recording**.

- Videos will be automatically saved to **Gallery App**.

- Press **Left Menu Button** to preview.

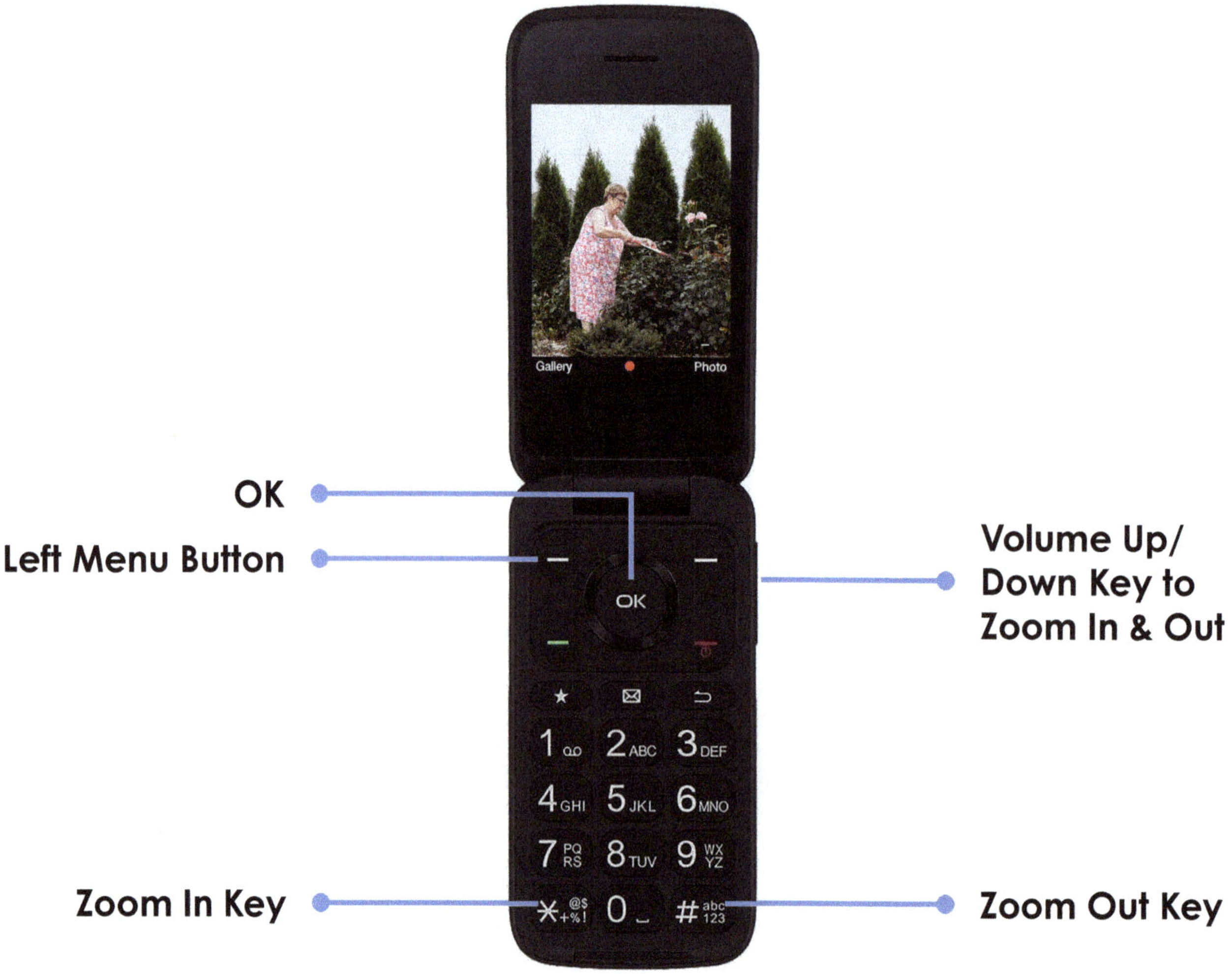

OK

Left Menu Button

Volume Up/Down Key to Zoom In & Out

Zoom In Key

Zoom Out Key

Accessing Photos and Videos – TCL Phones

7 **Previewing your Photos and Videos from the Camera App**

- From the Camera screen in **Photo Mode** or **Video Mode**, press the **Left Menu Button** to select **Gallery**.

Left Menu Button

8 Previewing your Photos and Videos from the Apps Screen

- From the **Home Screen**, press **OK** to go to the **Apps Screen**.

- Use the **Navigation Ring** to select the **Gallery App**  and press **OK**.

Gallery App

Apps Screen

OK

Navigation Ring

TCL Flip Phone

128

9 Finding your Photos and Videos in the Gallery App

- The **Gallery App**  provides you with one location to view and organize all your saved photos and videos.

- Use the **Navigation Ring** to select a photo or video and press **OK** to view.

- Press the **Right Menu Button** to access more **Options** like Delete, Share, Select Files, Edit, Set as wallpaper, Set as contact photo, and View favorites.

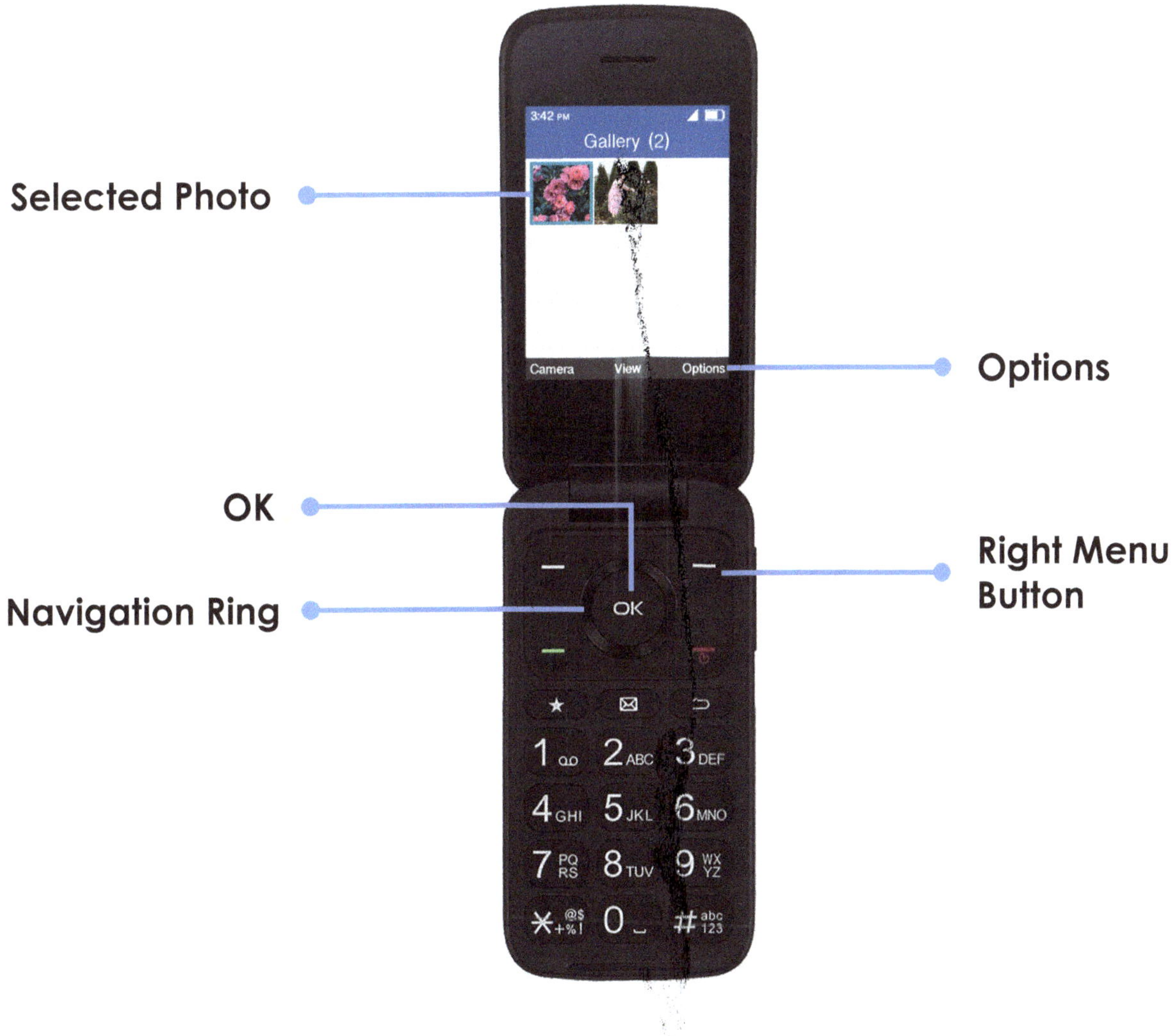

Selected Photo

Options

OK

Right Menu Button

Navigation Ring

Taking Pictures – iPhones

In this hands-on training, we'll walk you through the process of taking pictures and selfies of your family and loved ones. By the end of this guide, you will be more comfortable and confident taking pictures and selfies.

 **Unlocking Your Smartphone**

- Press the **Power/Side Button** on your Smartphone to wake it up.
- If your phone has a security feature like a **PIN**, **Pattern ID**, **Touch ID** (fingerprint), or **Face ID** lock, unlock it by following the on-screen prompts.

Power/Side Button

2 Accessing the Camera App

- Look for the **Camera App Icon** on your home screen or look for it in your **App Drawer**.

- Tap the **Camera App**  on the **Home Screen** or touch and hold the **Camera Icon** on the iPhone Lock Screen.

Camera App Icon

App Drawer

③ Taking a Picture

- Position the object or landscape in the screen, and tap the white **Shutter Button** to take the shot.

Thumbnail Image

Shutter Button

4 Zooming In & Out

- Open the **Camera App**  and pinch thumb and pointer finger together touching the screen and it will zoom in. Do the opposite to zoom out.

- This is the most universal method of zooming in or out.

- You may also tap one of the **Zoom Buttons** to instantly zoom at that zoom level.

Shooting Videos – iPhones

5 **Switching to Video Mode**

- Open the **Camera App** , then swipe right on the screen to select **Video Mode** or tap **Video**.

- For your security, a green dot appears at the top of the screen signaling **Camera In Use**.

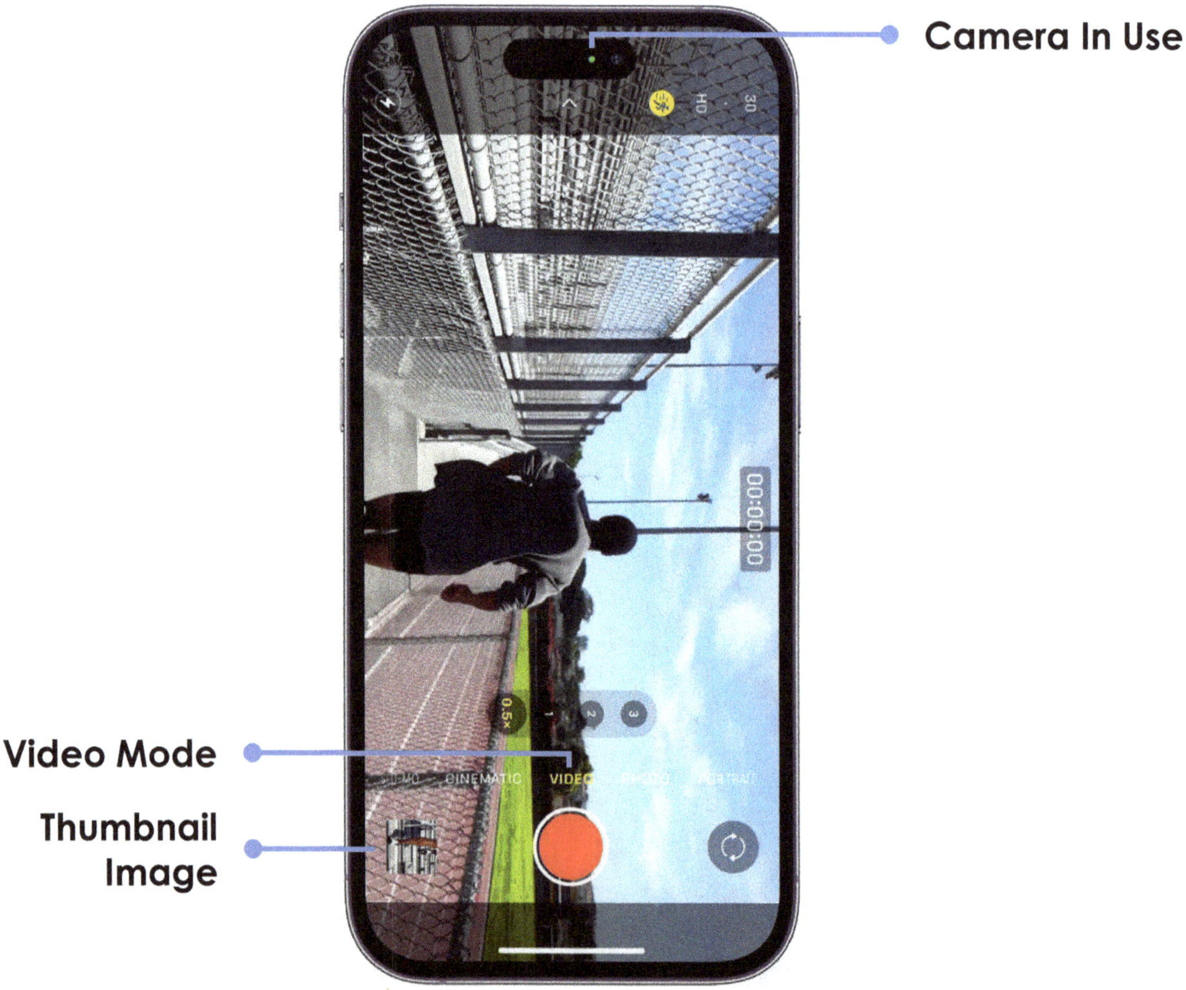

Camera In Use

Video Mode

Thumbnail Image

iPhone

 **Shooting a Video**

- Tap the red **Record Button** to start recording or press either **Volume Key** to start recording.

- Pinch thumb and pointer finger together touching the screen to zoom in. Do the opposite to zoom out.

- Tap the red **Record Button** to stop recording or press either **Volume Key** to stop recording.

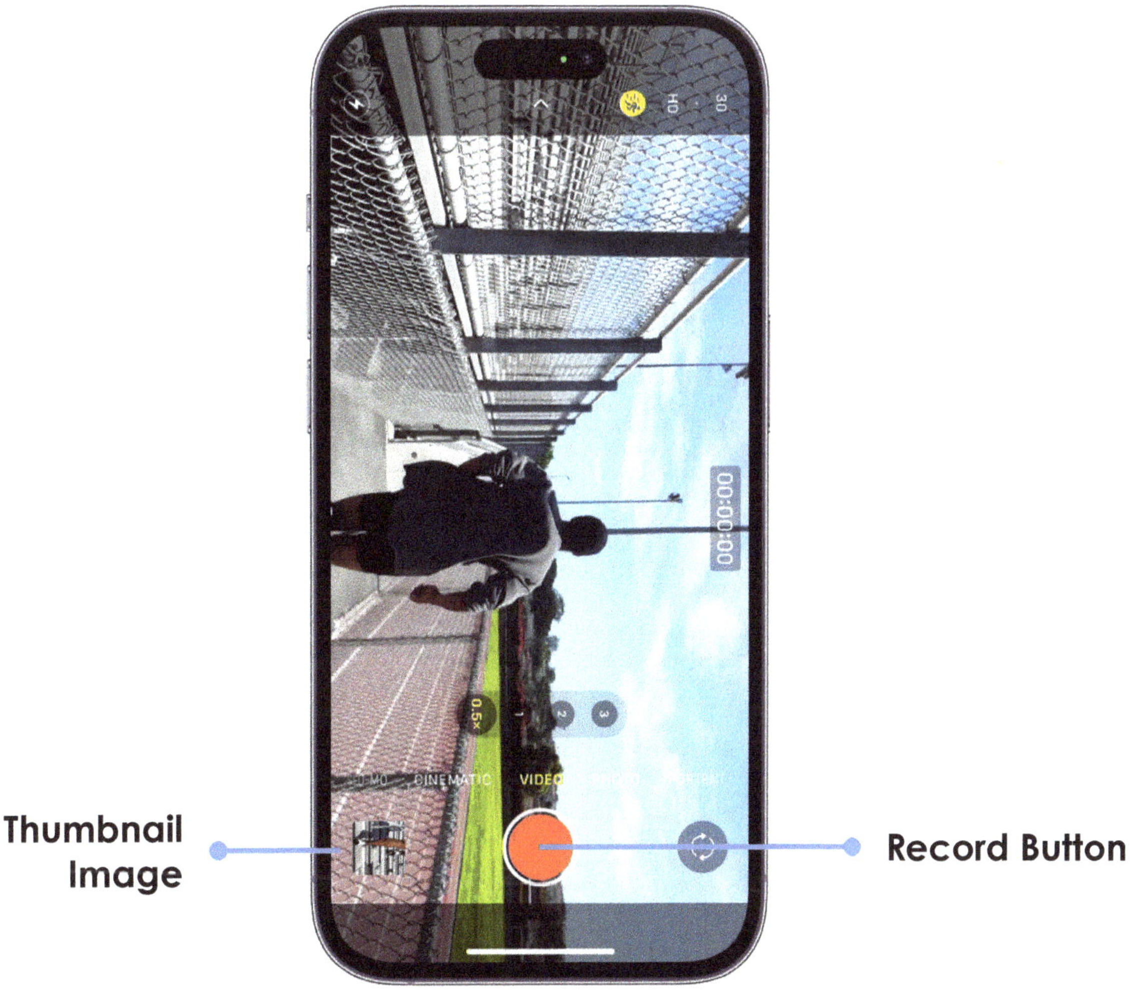

Thumbnail Image

Record Button

iPhone

Accessing Photos and Videos – iPhones

7. Previewing Your Photos and Videos from the Camera App

- Open the **Camera App** , then tap the **Thumbnail Image** in the lower-left corner.

- Swipe left or right to see the photos and videos you've taken recently.

- Tap the screen to show or hide the **Controls**.

- Tap **All Photos** to see all your photos and videos saved in Photos.

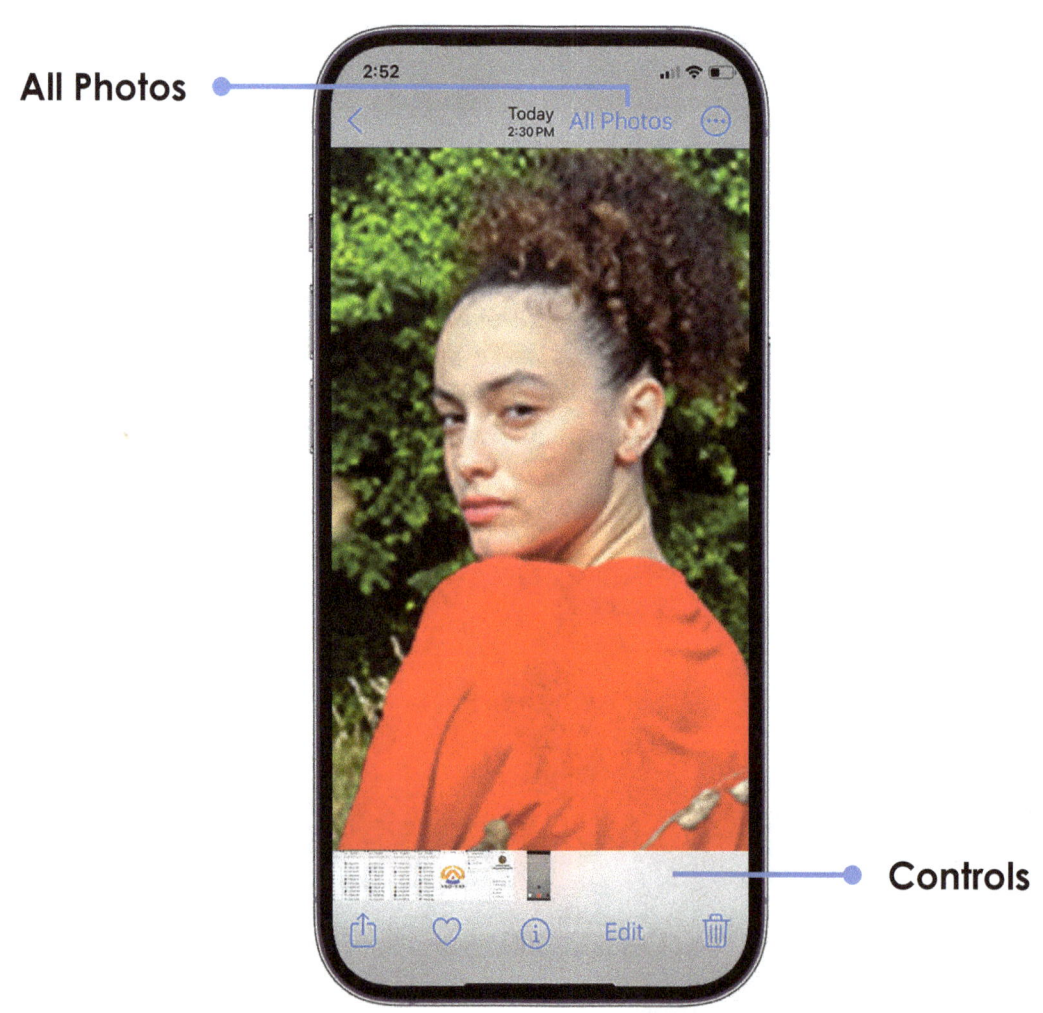

All Photos

Controls

8 Previewing Your Photos and Videos from the Photos App

- From the **Home Screen**, tap the **Photos App** to find and view all of the photos and videos on your iPhone.

- Swipe up and down to browse your photos and videos that are organized by **Years**, **Months**, **Days**, and **All Photos**.

- Tap a photo to view it in full screen on your iPhone.

- Tap the **Favorite (Heart) Icon** to add the photo to your Favorites album.

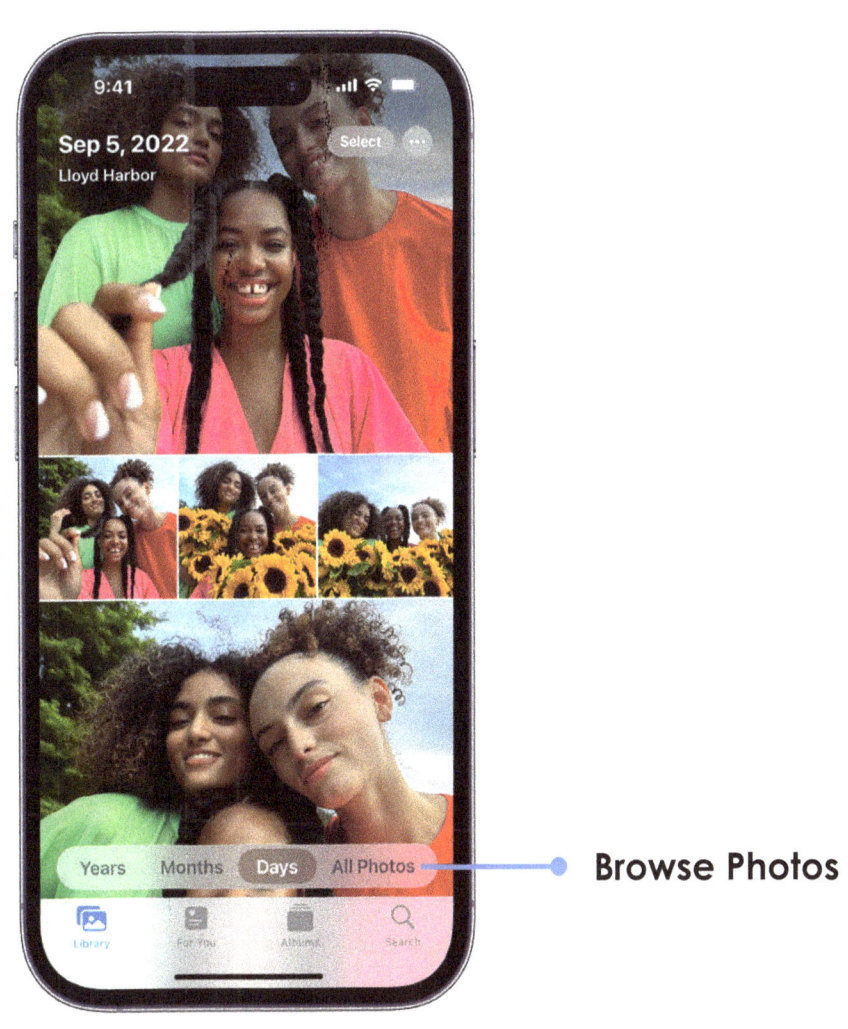

Browse Photos

137

Taking Selfies – iPhones

9 **Taking a Selfie with Your iPhone camera**

- Open the **Camera App** , then tap the **Switch Camera Icon** to switch to the front-facing camera.

- Hold your iPhone in front of you.

- Tap the white **Shutter Button** or press either **Volume Key** to take a selfie or start recording a video.

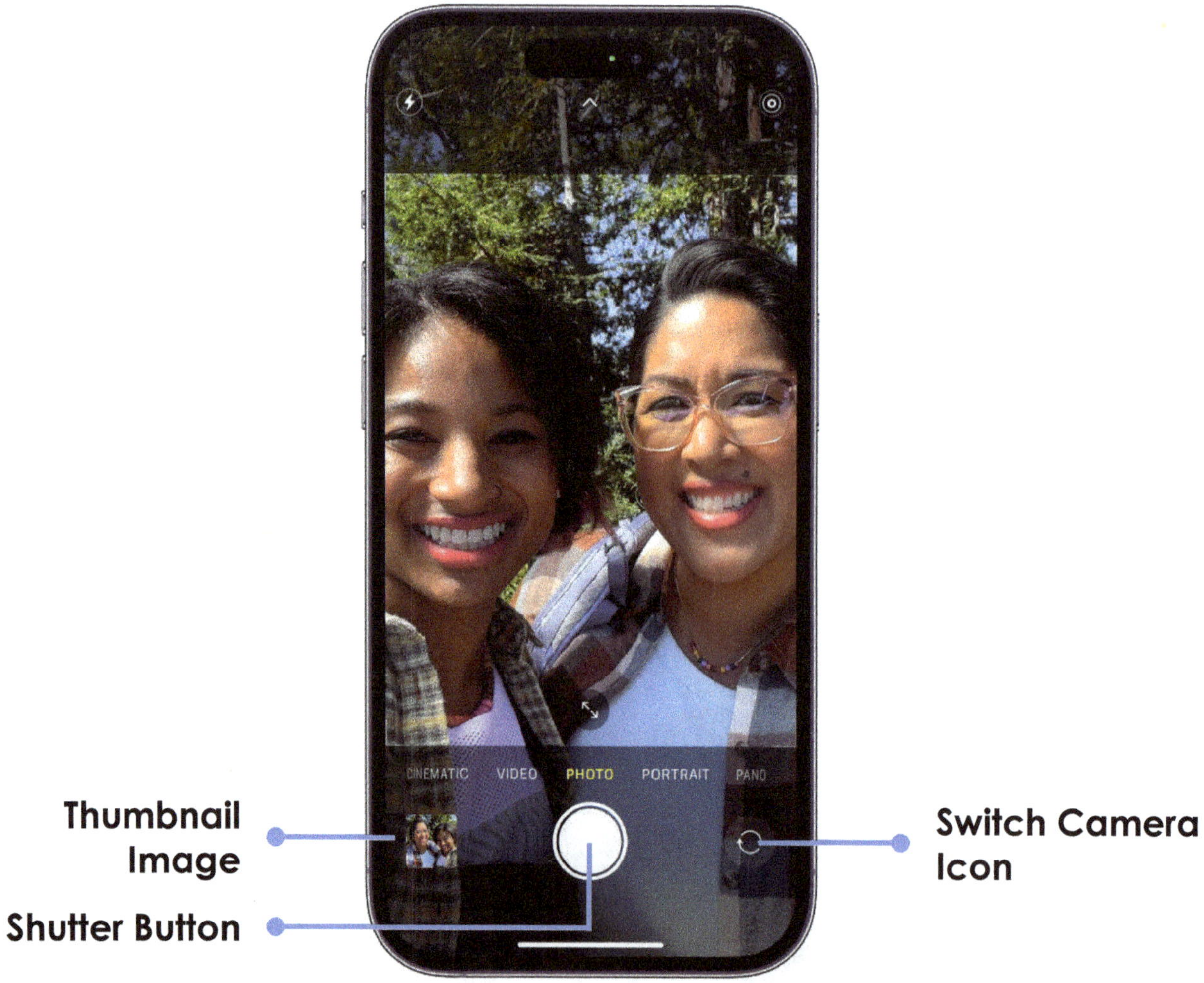

Thumbnail Image

Shutter Button

Switch Camera Icon

138

Taking Pictures – Galaxy Phones

In this hands-on training, we'll walk you through the process of taking pictures and selfies of your family and loved ones. By the end of this guide, you will be more comfortable and confident taking pictures and selfies.

1 **Unlocking Your Smartphone**

- Press the **Power/Side Button** on your smartphone to wake it up.
- If your phone has a security feature like a **PIN**, **Pattern ID**, **Touch ID** (fingerprint), or **Face ID** lock, unlock it by following the on-screen prompts.

Power/Side Button

2 Accessing the Camera App

- Look for the **Camera App Icon** on your home screen. It often looks like a camera icon or camera symbol.

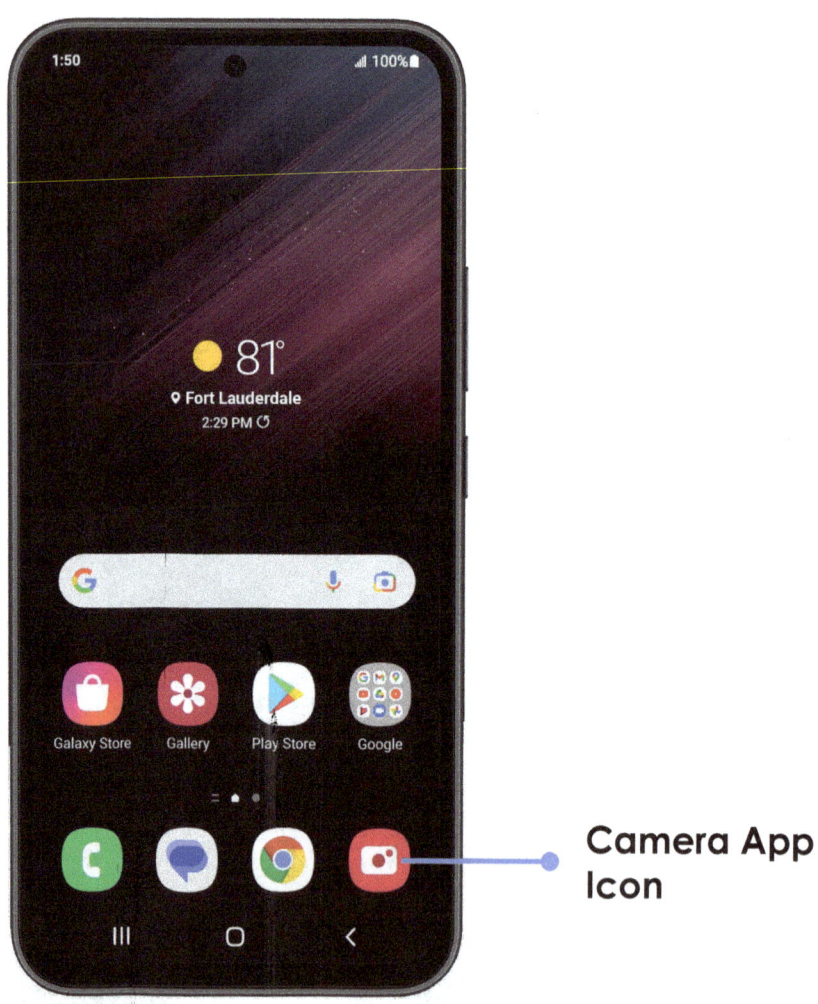

Camera App Icon

3 Taking a Photo

- Position the object or landscape in the screen, and tap the white **Shutter Button** to take the shot.

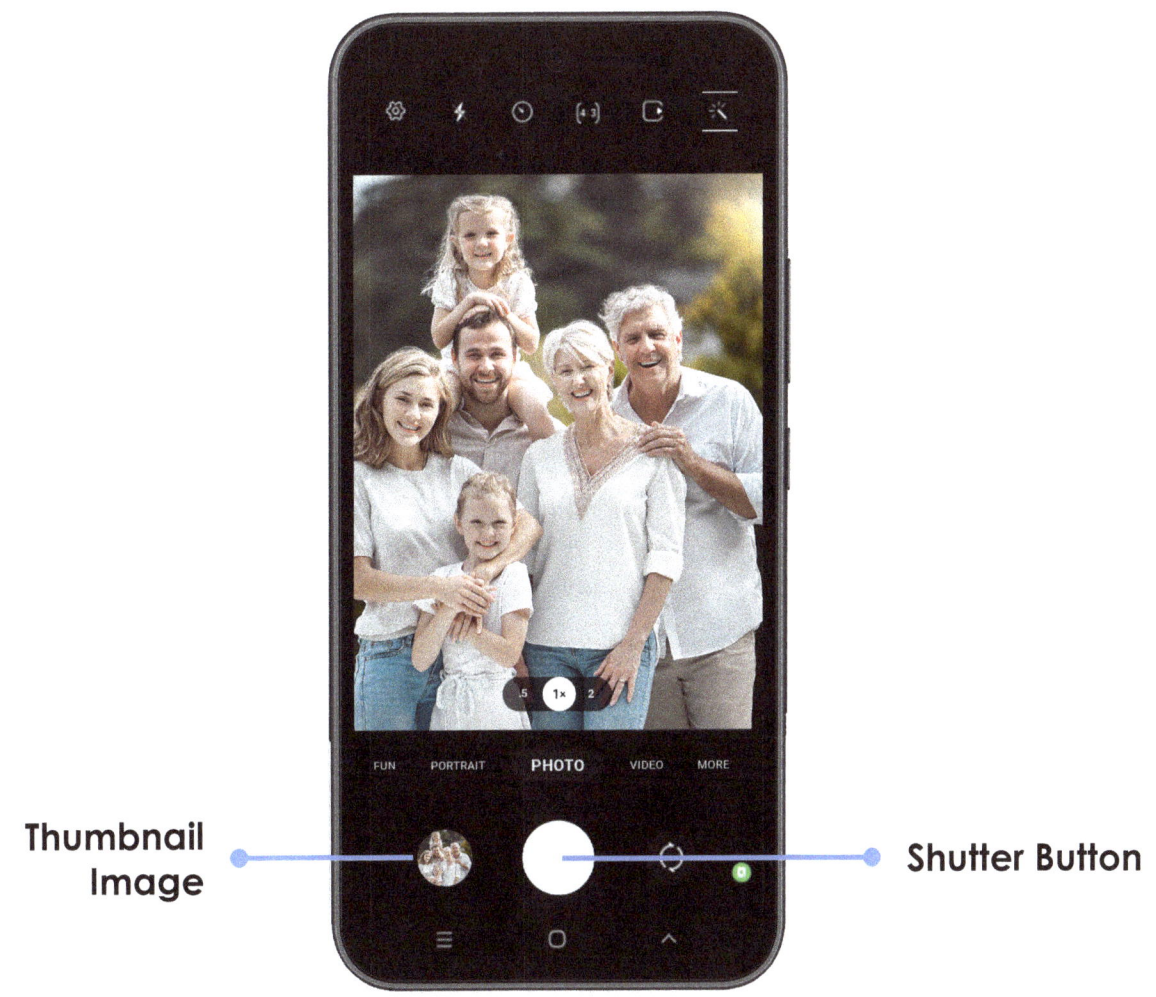

Thumbnail Image

Shutter Button

4 Zooming In & Out

- Open the **Camera App**  and pinch thumb and pointer finger together touching the screen and it will zoom in. Do the opposite to zoom out.

- This is the most universal method of zooming in or out.

- You may also tap one of the **Zoom Buttons** to instantly zoom at that zoom level.

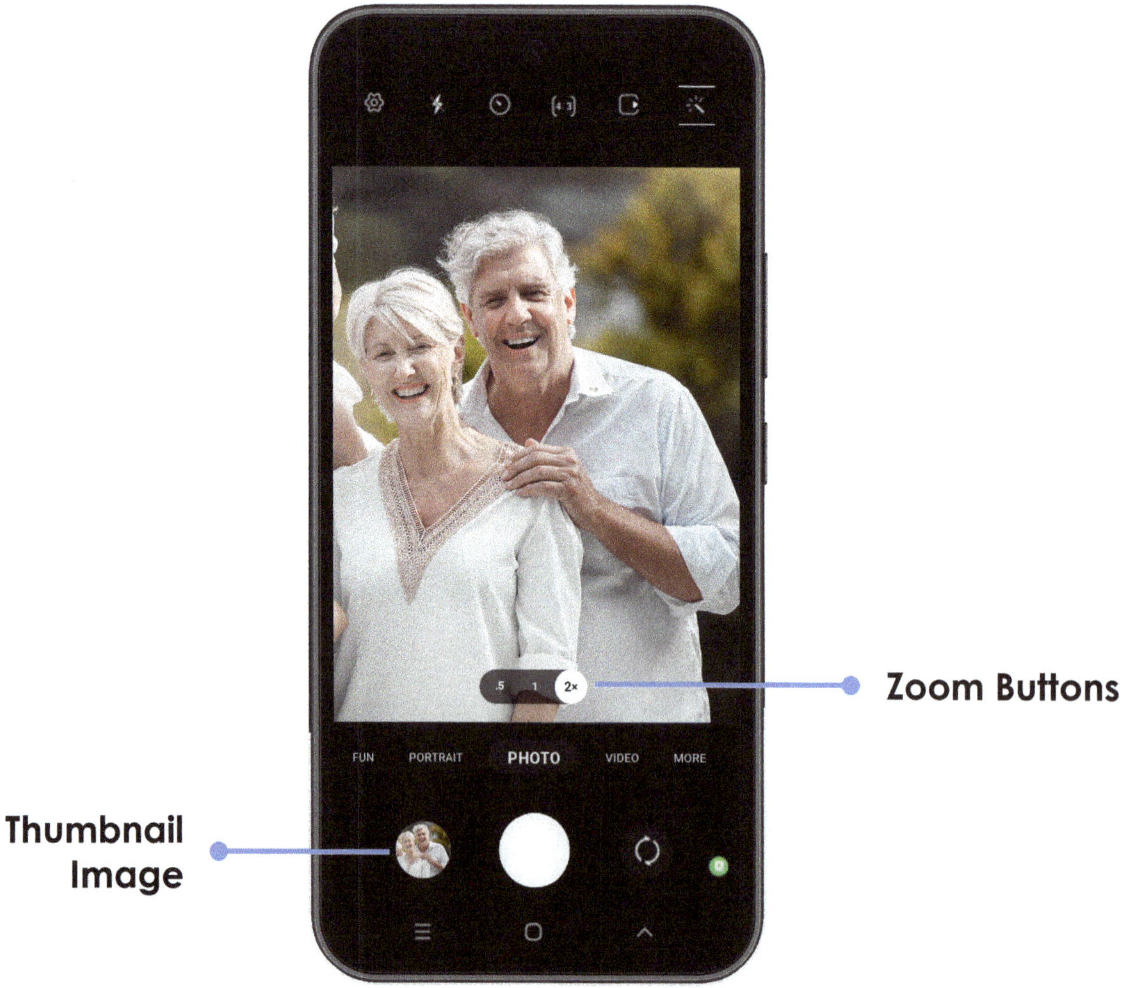

Zoom Buttons

Thumbnail Image

Shooting Videos – Galaxy Phones

5 Switching to Video Mode

- Open the **Camera App** , then tap **Video** on the **Camera Menu Bar** to select **Video Mode**.

Camera Menu Bar

Shooting a Video

- Tap the red **Record Icon** to start recording.
- Tap the **Stop Icon** to stop recording.
- Your videos are saved in the **Gallery App** along with your photos.

Record Icon

Accessing Photos and Videos

7 **Previewing Your Photos and Videos from the Camera App**

- Open the **Camera App** , then tap the **Thumbnail Image** in the lower-left corner.
- Swipe left or right to see the photos and videos you've taken recently.

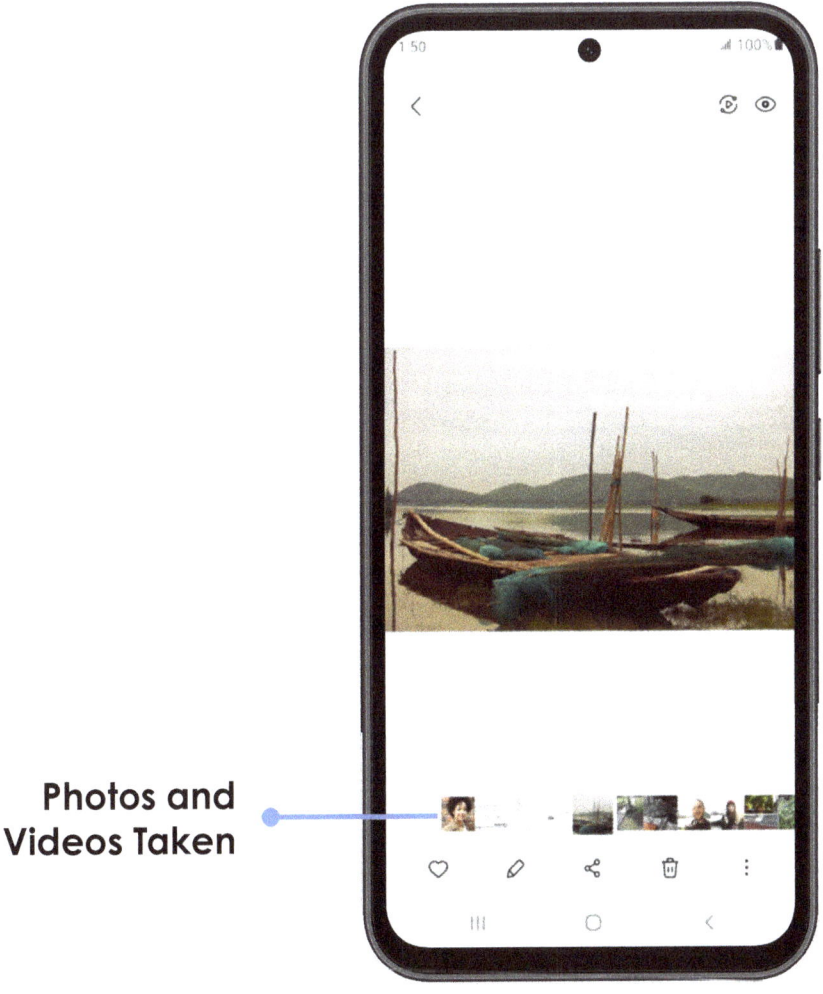

Photos and Videos Taken

8 Previewing Your Photos and Videos from the Gallery App

- From the **Home Screen**, tap the **Gallery App** ✿ to find and view all of the photos and videos on your iPhone.

- Swipe up and down to browse your photos and videos that are organized by date.

- Tap a photo to view it in full screen on your Galaxy smartphone.

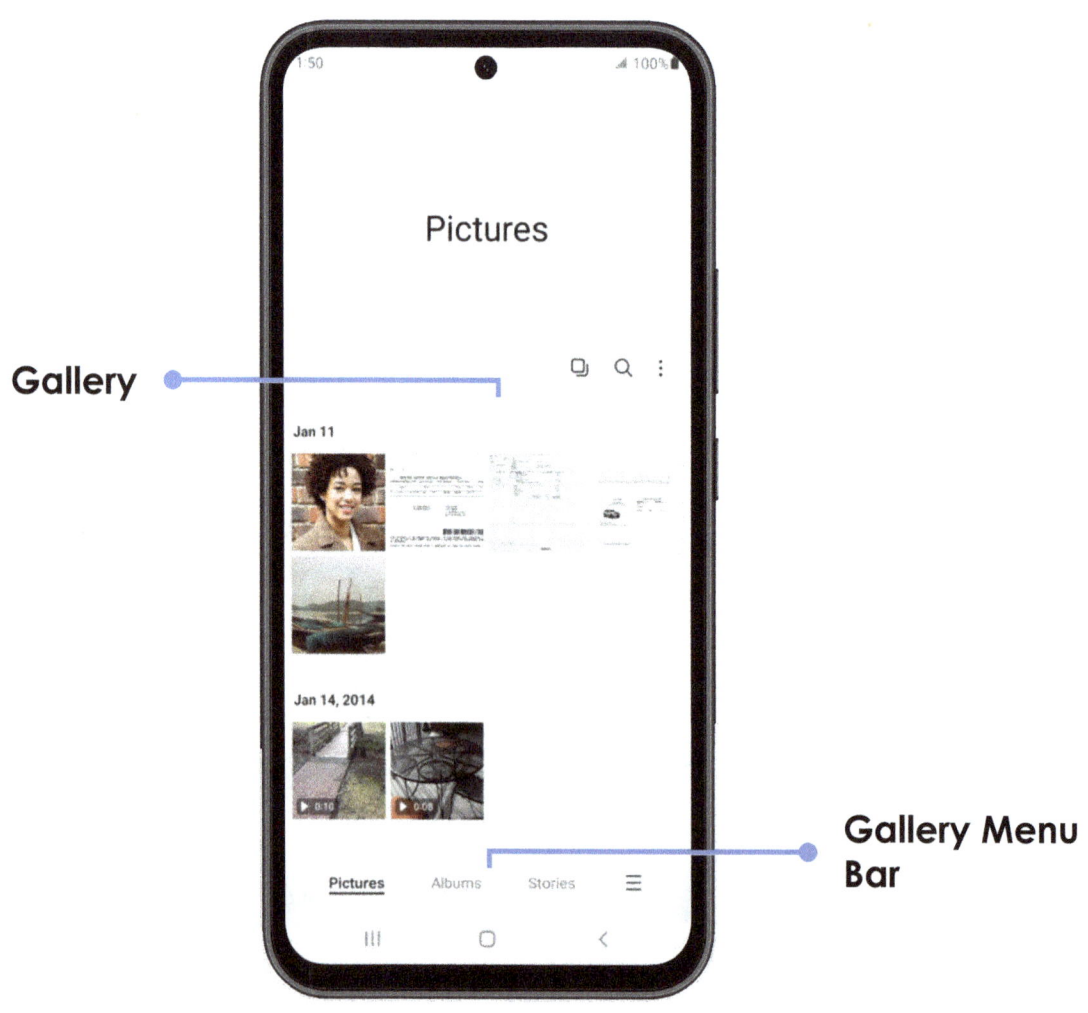

Galaxy Phone

146

Taking Selfies – Galaxy Smartphones

9 Taking a Selfie with Your Galaxy Smartphone Camera

- Open the **Camera App** 📷, then tap the **Switch Camera Icon** to switch to the front-facing camera.
- Hold your Galaxy smartphone in front of you.
- Tap the white **Shutter Button** to take a selfie.

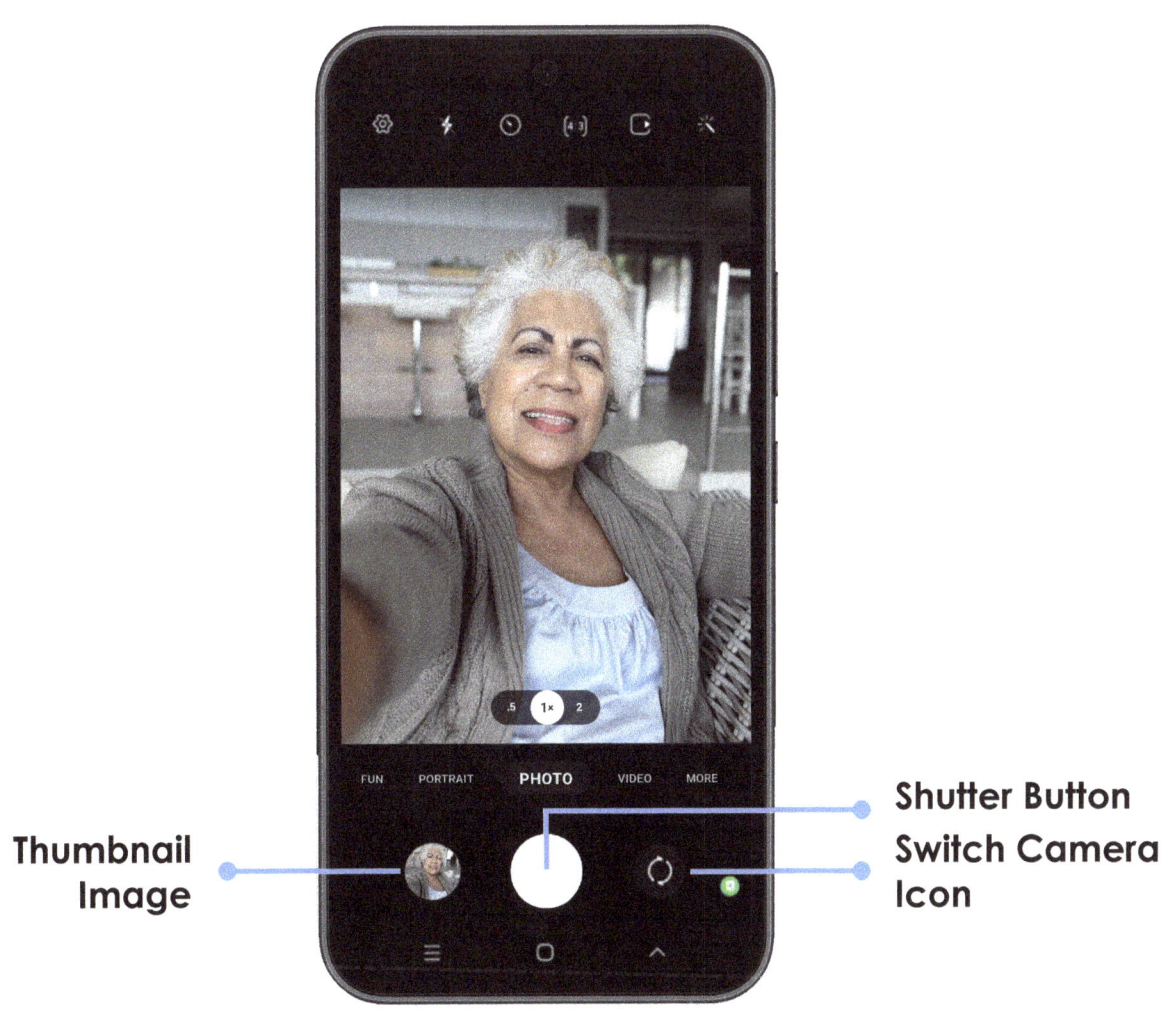

Chapter Questions

1. Taking a selfie means taking a picture of yourself.

    ○ True

    ○ False

2. To shoot a video, on an IPhone and Galaxy Phones, tap the Blue record button.

    ○ True

    ○ False

3. The camera app on the Galaxy Phone looks like a camera icon or camera symbol.

    ○ True

    ○ False

4. On the TCL Flip Phone when you shoot a video it will be automatically saved to Gallery App.

    ○ True

    ○ False

ANSWERS: 1. TRUE | 2. FALSE - Tap the Red button | 3. TRUE | 4. TRUE

BONUS – QR Codes

- A **QR code** is a type of matrix bar code that a digital device can scan. Most smartphones can scan QR codes with the **Camera App**.

- Scanning the **QR code** could take you to a website, social media account, online menu, event registration, sharing contact information, collecting payment, and playing a soundclip or music.

- Open the **Camera App** and hold your smartphone so that the **QR code** appears on the screen.

- Tap the notification that appears to open the link associated with the **QR code**.

- Now, let's do it! Scan the QR code below to open our website to learn more about **Senior Living Consultants**.

PREDICT 2 PREVENT™

Rm2.ai™ is an early warning platform that may predict potential health trends early to help reduce unnecessary hospital visits using continuous monitoring of vitals, and may provide support for proactive care to improve staff efficiency.

Meet Remy™

Your 24/7
Personal AI Assistant Companion™

SCAN HERE

Rm2.ai™ is an assistive technology platform designed to support proactive care, improve connectivity, and help reduce preventable complications. Rm2.ai™ is not a medical device and does not provide medical diagnosis, treatment, or guarantees of prevention. Outcomes may vary depending on individual circumstances, clinical oversight, and system usage. Rm2.ai™ should be used as a supportive tool alongside, not as a replacement for, professional medical judgment and care.

CHAPTER 5

Cell Phone Security and Troubleshooting

Addressing Security and Common Troubleshooting Items

Getting Familiar with Your Cell Phone's Security Features

Securing the Phone with Passwords or Biometrics

Securing your phone is crucial to protecting your personal data and maintaining privacy. There are several ways to achieve this:

Setting Up Your Password/PIN – TCL Phones

1 ### Phone Settings

- Open the **Settings** screen by pressing **OK** from the **Home Screen**.
- Use the **Navigation Ring** to highlight **Phone Settings**, then press **OK**.

Settings Screen

OK

Navigation Ring

2 Security

- Use the **Navigation Ring** to highlight **Security**, then press **OK**.

3 Setting Your Screen Lock

- Use the **Navigation Ring** to highlight **Screen Lock**, then press **OK**.

- Use the **Navigation Ring** to highlight **On**, then press **OK**.

4 Setting Your Passcode/PIN

- Use the **Keypad** to enter the desired **4-digit passcode** in the **Passcode Field**, then press **OK** to select **Next**. Enter the **passcode** again to confirm. When finished, press **OK** again.

- To turn off screen lock, from the **Security** screen, use the **Navigation Ring** to highlight **Screen Lock**, then press **OK**. Enter the current **4-digit passcode**, then press **OK** again. Move down to highlight **Off**, then press **OK** to select.

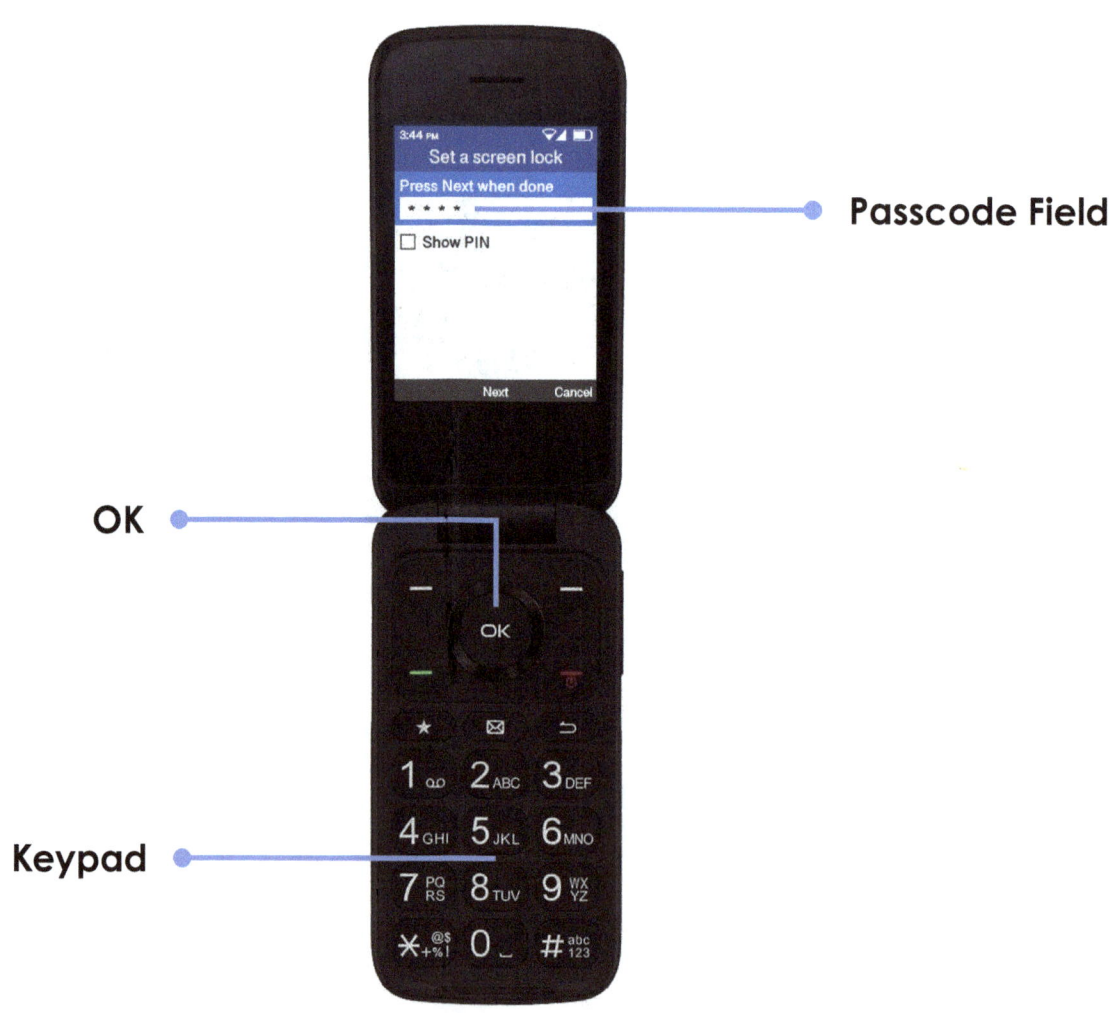

Passcode Field

OK

Keypad

Setting Up Your Password/PIN – iPhones

1 ### Setting Your Password/PIN

- Create a strong, unique password or PIN for your phone. Avoid easily guessable combinations like birthdays or simple patterns.

- To avoid errors, follow the password and PIN set up instructions from your phone manufacturer.

2 Using Biometrics (Facial/Face ID)

- Use facial recognition for added security. Ensure that only your face can unlock your device.

- Use Face ID to securely and conveniently unlock iPhone, authorize purchases and payments, and sign in to many third-party apps by simply glancing at your iPhone.

- To use Face ID, you must also set a passcode on your iPhone.

- Go to **Settings**, select **Face ID & Passcode**, select **Set up Face ID**, then follow the **Onscreen Instructions**.

3 Regularly Update Security Settings

- Stay updated with the latest security patches and updates for your phone's operating system (iOS or Android). Any available update will appear on the **Software Update Screen**. Operating System updates often include security enhancements.

- Follow the instructions listed in each update to begin.

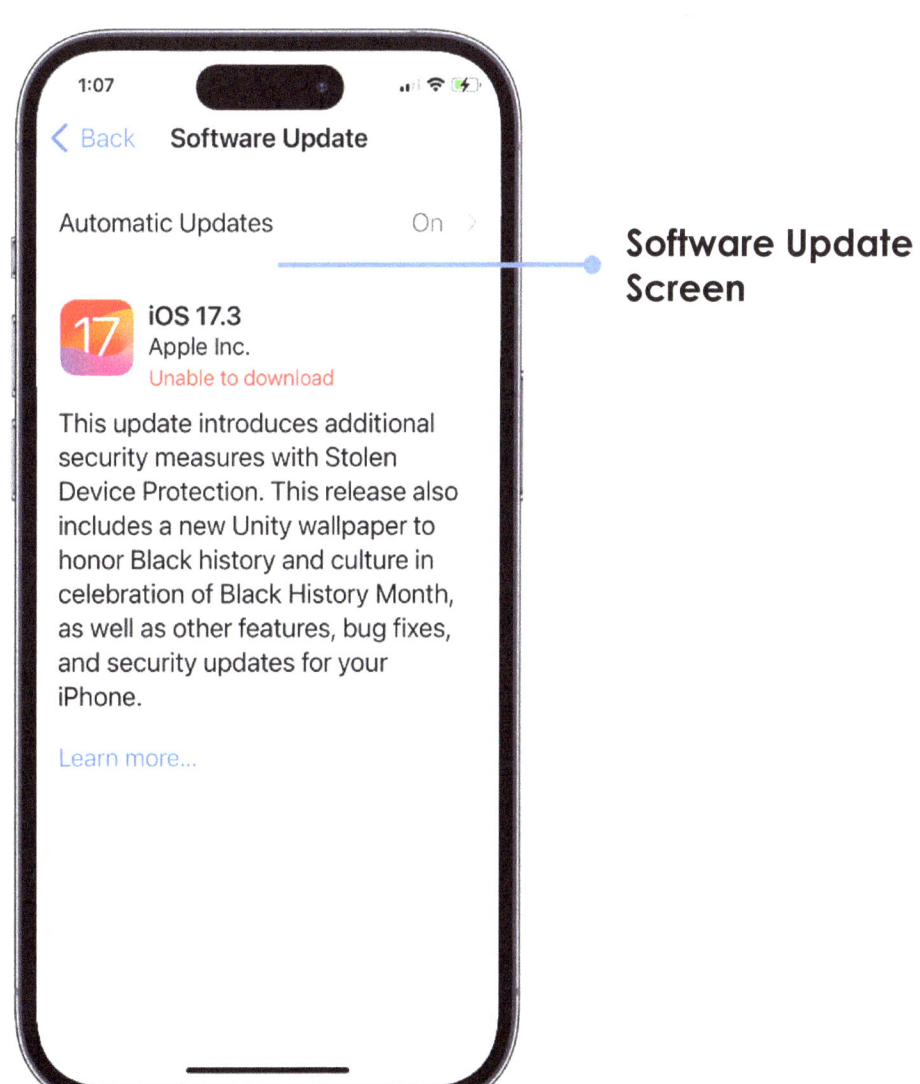

Software Update Screen

iPhone

159

Setting Up Your Password/PIN – Galaxy Phones

1 ### Setting Your Password/PIN

- Open the **Settings App** ⚙, then scroll to **Lock Screen** and tap it.

- Tap **Screen Lock Type** and tap **PIN**.

- Enter your **PIN** and tap **Continue**.

- Confirm your **PIN**, then tap **OK**.

- Tap **Hide Content** to prevent the content of emails, text, etc., then tap **Done**.

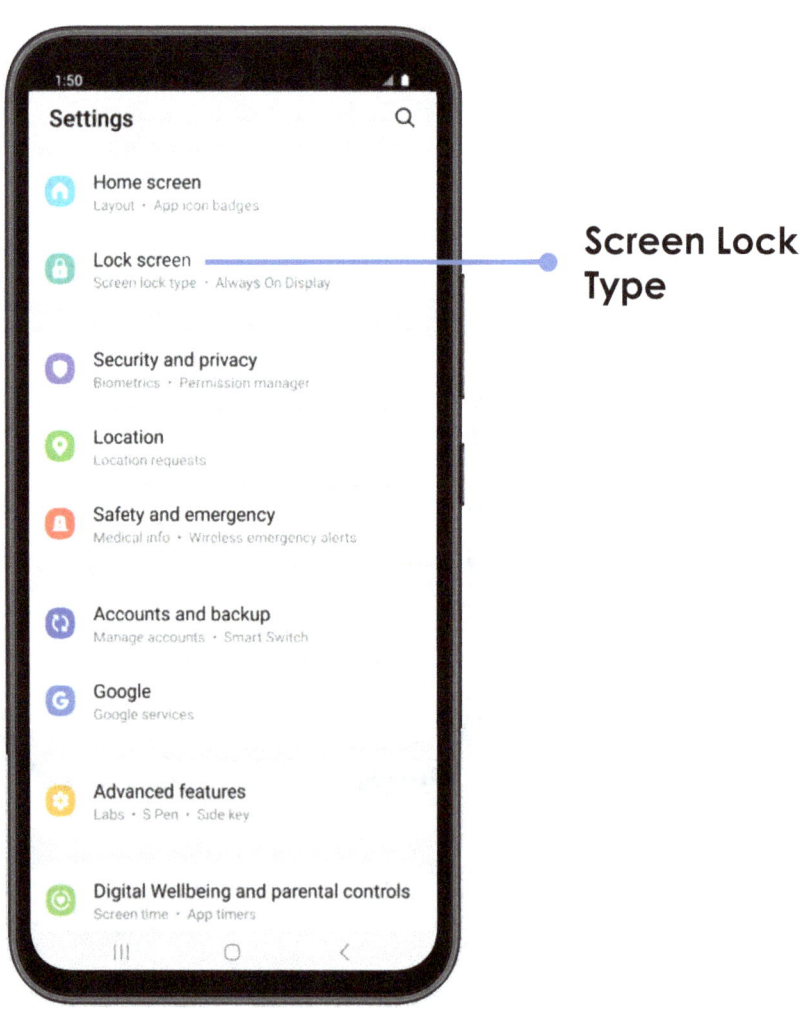

Screen Lock Type

2 Using Biometrics (Fingerprint Recognition)

- Fingerprint recognition uses the unique characteristics of your fingerprints to enhance the security of your device.

- Using a fingerprint to unlock your phone is faster and more convenient than using a PIN or password.

- Go to **Settings**, tap **Security and Privacy**, then tap **Biometrics**, and select **Fingerprints**.

- Tap the **Fingerprint Unlock Switch** to activate it.

- On the **Lock Screen**, place your finger on the **Fingerprint Recognition Sensor** and scan your fingerprint.

- Tap **Show Icon When Screen is Off** and select **On Always**.

Fingerprint Recognition Scanner

161

3 Regularly Update Security Settings

- Stay updated with the latest security patches and updates for your phone's operating system (iOS or Android). Any available update will appear on the **Software Update Screen**. Operating System updates often include security enhancements.

- Follow the instructions listed in each update to begin.

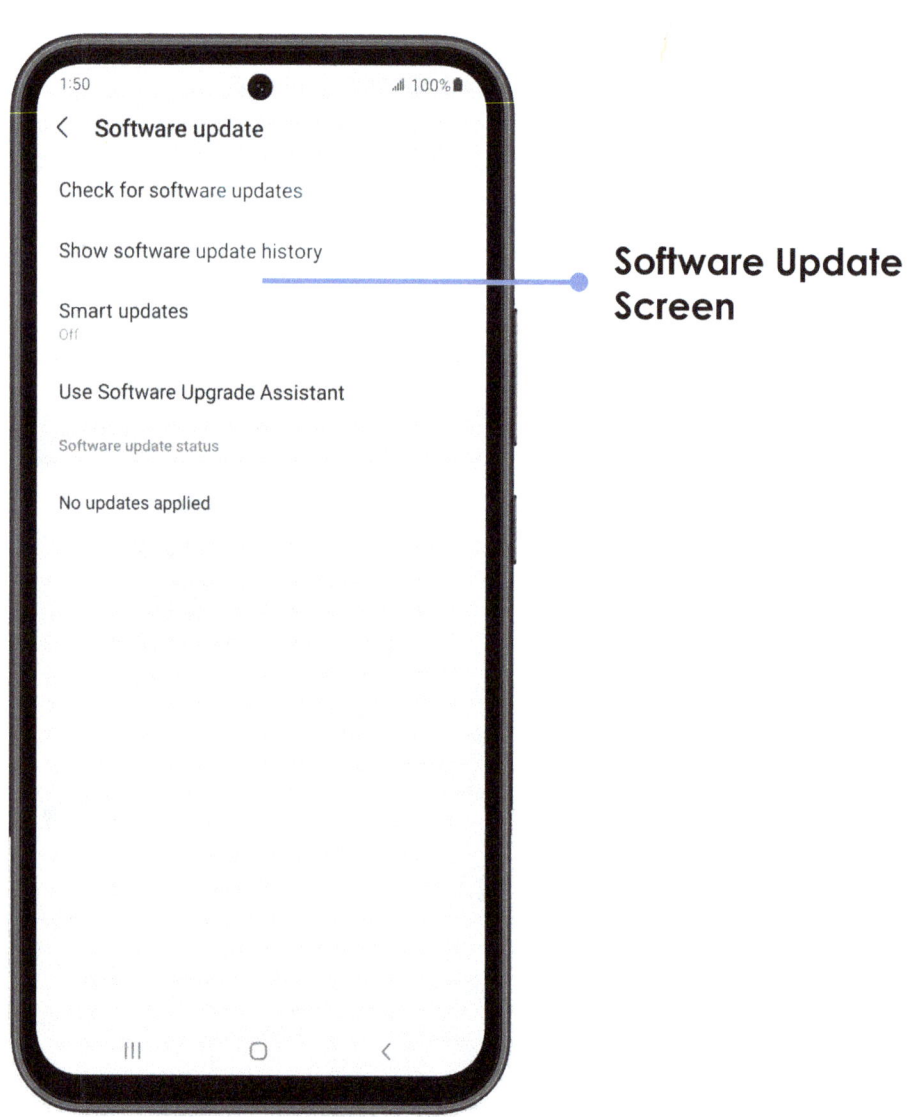

Software Update Screen

Protecting Personal Data and Recognizing Scams

Protecting personal data and being cautious about scams are essential for online safety.

1 ## Be Cautious with Personal Information

Avoid sharing sensitive personal information on suspicious websites or with unknown contacts. Be mindful of what you share on social media.

2 ## Use Trusted Apps and Websites

Download apps and visit websites from reputable sources. Verify the legitimacy of apps before downloading them. Legitimate apps typically do not have typos or grammatical errors and tend to have millions of downloads.

3 ## Educate Yourself About Scams

Stay informed about common scams, phishing attempts, and fraudulent activities. Be skeptical of unsolicited communications asking for personal information or financial details.

4 ## Use Anti-Malware Software

Install reputable anti-malware and antivirus software on your phone to protect against malicious apps and files.

5 Regularly Review Permissions

Check and manage app permissions on your phone to ensure that apps only have access to the information they truly need.

6 Beware of Phishing Attempts

Be cautious of emails, messages, or calls asking for sensitive information. Verify the sender's identity, and never click on suspicious links.

Chapter Questions

1. Setting a pin ensures maximum security for a cell phone.

    ○ True
    ○ False

2. Regularly updating security settings never includes security enhancements

    ○ True
    ○ False

3. Anti-Malware software on your phone helps protect against malicious apps and files.

    ○ True
    ○ False

ANSWERS: 1. TRUE | 2. FALSE | 3. TRUE

CHAPTER 6

Real-Time Text (RTT) and Teletype (TTY)

Assisting Seniors with Hearing and Speech Impairments to Communicate by Cell Phone

RTT and TTY – iPhones

This feature allows seniors or anyone with hearing or speech impairments to communicate by cell phone using Real-Time Text (RTT) or Teletype (TTY).

Set up and use RTT and TTY

- Go to **Settings**, then tap Accessibility.

- Tap **RTT/TTY**, turn on **Software RTT**, then tap **Relay Number**, then enter the phone number to use for relay calls using Software RTT/TTY.

- Turn on **Send Immediately** to send each character as you type. Turn it off to complete messages before sending.

- Turn on **Answer All Calls** as RTT/TTY.

- If you want to answer and make RTT/TTY calls from an external RTT/TTY device instead of your iPhone, turn on **Hardware TTY**.

- To start an RTT or TTY call open the **Phone App**, tap **Contacts** and choose a contact, then tap the **Phone Number**.

- Choose **RTT/TTY Call** or **RTT/TTY Relay Call**, then wait for the call to connect, then tap the **RTT Button**.

- To type text during an RTT or TTY call tap the **RTT Button**.

- Enter your message in the text field.

- To also transmit audio, tap the **Microphone Off Muted Button** in the top-right corner.

RTT and TTY – Galaxy Phones

This feature allows seniors or anyone having hearing or speech difficulties to communicate effectively over the cell phone using Real-Time Text (RTT) or Teletype (TTY).

Adjusting RTT and TTY Call Settings

- Go to the **Phone App**, tap the **Menu Icon** (three vertical dots), then tap **Settings**.

- Scroll to and tap **Other Call Settings**, tap **TTY Mode**, then choose preferred TTY setting and tap the **Back Icon**.

- Scroll to and tap **Real Time Text**, tap **RTT Call Button**, then choose when you would like the RTT Call Button to appear.

- To start an RTT or TTY call open the **Phone App**, tap **Contacts** and choose a contact, then tap the **Phone Number**.

- Tap **RTT**. While the phone rings, the other person's screen displays an invitation to join the RTT call.

- After the other person answers, enter a message in the text field.

- When you're done with the call, tap **End Call**.

- To switch from voice to RTT during a call, tap **RTT**.

- RTT calls include an audio stream. To mute or unmute your microphone, tap **Mute**.

CHAPTER 7

The Benefits of Artificial Intelligence (AI) and Voice Activated Technology for Seniors

Providing Insights on How AI can make Tasks Easier and More Personalized

Artificial Intelligence Assistants

Artificial Intelligence (AI) can assist you in using your cell phone more effectively through:

1 ## Voice Commands
Enabling hands-free control with voice assistants, allowing seniors to make calls, send messages, and perform tasks without navigating complex menus.

2 ## Personalized Assistance
Tailoring recommendations and settings based on individual preferences, making the user interface more intuitive and accessible.

3 ## Speech-to-Text and Text-to-Speech
Converting spoken words to text and vice versa, aiding communication for those with hearing or speech difficulties.

4 ## Smartphone Accessibility Features
Leveraging AI to enhance built-in accessibility features, such as screen readers, magnification, and voiceover, catering to specific needs.

5 AI-Driven Tutorials

Providing step-by-step tutorials and guidance on using various phone functions, adapting to the user's pace and learning style.

6 Automated Assistance

Automating routine tasks through AI-driven routines, reducing the need for manual navigation and streamlining daily activities.

7 Predictive Text and Autocorrect

Offering predictive text suggestions and autocorrect features to help users type more accurately and efficiently.

8 Health Monitoring Apps

Integrating health monitoring apps that use AI to track vital signs, remind users of medication, and offer health-related information.

By incorporating these AI-driven features, you can experience a more user-friendly and personalized interaction with your cell phone, enhancing overall accessibility and usability.

Hands-free Voice Assistants

Instructions on how to use hands-free assistants on your smartphone:

1 ### Activate Voice Assistant

On iPhones, use Siri by saying "Hey Siri" or press and hold the side/home button.

On Galaxy phones, use Google Assistant by saying "Hey Google" or press and hold the home button.

2 ### Issue Commands

For tasks like sending messages, say "Send a message to [contact] ... " followed by your message.

To make calls, say "Call [contact]" or "Dial [number]."

3 ### Use Navigation

For directions, say "Navigate to [destination]" to open navigation apps.

4 ### Set Reminders

Say "Set a reminder for [task] at [time]" to schedule reminders.

5 Ask Questions

Pose questions like "What's the weather today?" or "Define [word]."

6 Control Music

Control music playback with commands like "Play [song]" or "Pause music."

7 Open Apps

Open apps by saying "Open [app name]."

Remember to enunciate clearly and experiment to discover the full range of voice assistant capabilities on your specific device.

Voice Activated Technology

Voice activation on cell phones can be a valuable tool for you to navigate your device more easily. Here are some tips for you to effectively use voice activation:

1 Familiarize Yourself with Voice Commands

Spend some time getting to know the voice commands available on your phone. Practice using common commands like "Call [Corey]," "Text [Alexis]."

2 Speak Clearly and Loudly

Ensure that you speak clearly and loudly when giving voice commands to your phone. Speaking directly into the microphone will help your phone accurately recognize your commands.

3 Use Natural Language

Many voice assistants understand natural language, so you don't have to use specific phrases. For example, instead of saying "Call Mom's mobile," you can say "Call my mom."

4 Start with Basic Commands

Begin with basic commands and gradually explore more advanced features as you become more comfortable with voice activation. This approach will help prevent frustration and make the learning process smoother.

5 Practice Regularly

Like any new skill, practice is key. Spend some time each day practicing voice commands until you feel comfortable and confident using them.

6 Adjust Settings

Some phones allow you to adjust voice activation settings to better suit your needs. For example, you may be able to adjust the sensitivity of the microphone or enable a feature that allows you to wake up the voice assistant with a specific phrase.

7 Explore Accessibility Features

Many smartphones offer accessibility features specifically designed for seniors and individuals with disabilities. These features may include voice control options tailored to specific needs.

8 Seek Assistance if Needed

Don't hesitate to ask for assistance from a family member, friend, or caregiver if you're having difficulty with voice activation. They can provide guidance and support as you learn to use this feature.

9 Stay Patient

Learning to use new technology can take time, so be patient with yourself as you explore voice activation on your cell phone. With practice and perseverance, you'll become more proficient over time.

10 Keep Learning

Stay curious and continue learning about the capabilities of voice activation on your cell phone. New features and updates are frequently released, so staying informed will help you make the most of this technology.

Senior Resources

Senior Living Consultants
www.seniorlc.com

National Institute on Aging
www.nia.nih.gov

Alzheimer's Association
www.alz.org

Meals On Wheels America
www.mealsonwheelsamerica.org

American Associations of Retired Persons
www.aarp.org

National Council On Aging
www.ncoa.org

Administration for Community Living
www.acl.gov

Social Security Administration
www.ssa.gov

Remember2
www.remember2.ai

Pushing The Right Button
CROSSWORD PUZZLE

Across
3. What do you compose and send through your phone?
5. What square barcodes are used for scanning with a smartphone camera?
7. What type of security method uses unique physical characteristics for identification?
8. What device allows you to make calls and send messages?
10. When you call 911 what kind of situation is it?
12. When receiving an incoming call your phone may ring or _____.
13. On the IPhone, What button do you tap to reject a a call when your phone is unlocked?
14. What icon do you tap to switch between active calls?

Down
1. What do you do to keep your personal information safe from hackers?
2. What feature allows seniors or anyone with hearing or speck impairment to communicate by cell phone?
4. What type of photo do you take of yourself using your phone's camera?
6. What type of communication is sent through your phone in written form?
9. What do you do to ensure your data is safe from harm?
11. What button do you tap when you are finished with a call?

Pushing The Right Button
WORD SCRAMBLE

1. HMNTEACSTAT _____
2. RBEIVTA _____
3. OHPOST _____
4. TBUTNO _____
5. APOESDSC _____
6. LFISEE _____
7. MSOAMCDN _____
8. NEIECLD _____
9. ACMARE _____
10. SCAOIL ADMEI _____
11. NTNLIILGCEEE _____
12. KPAYDE _____
13. TGVNAAONII _____
14. NEIOPH _____
15. ECRYNMEEG _____
16. TISECRUY _____
17. OLECTOHGNY _____
18. FILICIATRA _____
19. SUMNASG _____
20. OIESDV _____
21. PLAPE _____
22. XLAGYA _____
23. NIMZOOG _____
24. NSPAMEOSRHT _____
25. INENTRET _____

Word Bank

Selfie	Button	Samsung	Passcode	Security
Technology	Zooming	Camera	Photos	Videos
Emergency	Internet	Keypad	Commands	Artificial
Galaxy	Social Media	iPhone	Vibrate	Navigation
Decline	Attachments	Intelligence	Smartphones	Apple

Answer Key

Crossword Puzzle – page 178

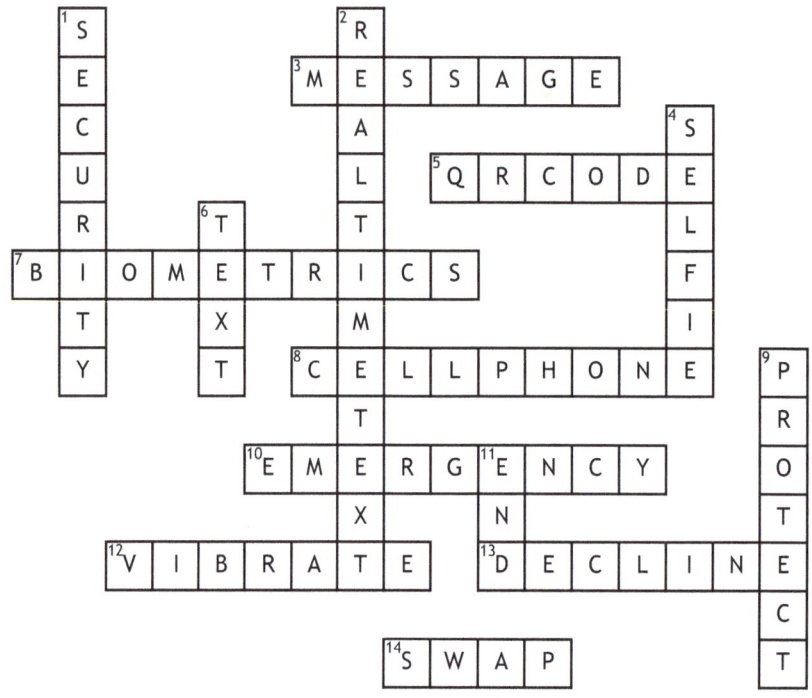

Word Scramble – page 179

1. **HMNTEACSTAT** Attachments
2. **RBEIVTA** Vibrate
3. **OHPOST** Photos
4. **TBUTNO** Button
5. **APOESDSC** Passcode
6. **LFISEE** Selfie
7. **MSOAMCDN** Commands
8. **NEIECLD** Decline
9. **ACMARE** Camera
10. **SCAOIL ADMEI** Social Media
11. **NTNLIILGCEEE** Intelligence
12. **KPAYDE** Keypad
13. **TGVNAAONII** Navigation
14. **NEIOPH** iPhone
15. **ECRYNMEEG** Emergency
16. **TISECRUY** Security
17. **OLECTOHGNY** Technology
18. **FILICIATRA** Artificial
19. **SUMNASG** Samsung
20. **OIESDV** Videos
21. **PLAPE** Apple
22. **XLAGYA** Galaxy
23. **NIMZOOG** Zooming
24. **NSPAMEOSRHT** Smartphones
25. **INENTRET** Internet

Conclusion

Don't underestimate the advantages of mastering cell phone use. Learning methods vary, so embrace your own style without hindrance. Congratulations on taking the initiative to learn and create cherished memories with loved ones through using your cell phone. We trust this guide has been valuable in familiarizing you with the benefits of cell phones and enhancing your comfort in using them. Now, leverage the tools provided in this book to navigate your device confidently and effectively. Go ahead and *Push The Right Button*.

About The Author

Corlette Deveaux, MBA
CEO, Senior Living Consultant

Corlette Deveaux is a philanthropist, senior advocate, and a true humanitarian. Her volunteerism and advocacy spans over 25 years. She is a member of 2023 and 2024 Executive Committee for the Alzheimer's Association, Broward County Walk. She is a long and active member of the Rotary Club of Weston, vice president of her homeowner's association, former chairperson of Parkland Chamber of Commerce, vice president of the Board of Pink Pearls of Miramar Foundation, 2015 Woman of the Year with the National Association of Professional Women and a 2023 Recipient of an Global Trade Award.

Corlette has over 25 years of experience with large organizations such as Pfizer, Novartis Pharmaceuticals, Alcon Labs, and Bio-Tissue Inc in a number of capacities including sales, strategic marketing, and operations. She gained valuable experience educating and training medical professionals in various areas such as Alzheimer's disease, hypertension, diabetes, age-related macular degeneration, osteoporosis, allergies, and more. In her roles, she has maintained responsibilities for large brands valuing in excess of 500 million dollars. Her successful experience propelled her to the top of her profession with superior results that allowed her to help organizations build and exceed expectations.

She has leveraged her extensive corporate experience to forge ahead in her own entrepreneurial pursuits. In 2015, she refined her focus on the senior industry. In early 2016, she opened an adult day care center. She works strategically with clients in the senior industry to help them open, operate, and grow senior living facilities. She is a highly respected trainer, providing coaching and leadership training to large audiences worldwide.

Corlette continued focus is on building her reputation in the senior industry, providing training and education for clients, families, and caregivers. She is the host of an online show called *Our Aging Puzzle* which airs on

Fridays at 8 p.m. on her YouTube channel @dgincorporated and @ouragingpuzzle. Corlette was featured in a one-on-one interview with Kevin Harrington as part of American Entrepreneur and their consulting services. The full interview can be watched on her website, www.SeniorLC.com.

Her company, Senior Living Consultants, provides services to entrepreneurs who are interested in filling an unmet need in the senior industry. As a Florida licensed assisted living facilitator (ALF) administrator, she is familiar with the compliance and regulatory requirements in assisted living facilities. As a senior advocate and trainer, she conducts educational training workshops for seniors on various topics including fall prevention, maintaining memory, understanding technology, and more. She has devoted her life to helping enhance the lives of seniors.

CONTACT ME

Email:
contact@seniorlc.com

Website:
www.SeniorLC.com

www.ingramcontent.com/pod-product-compliance
Lightning Source LLC
Chambersburg PA
CBHW062130160426
43191CB00013B/2251